Plasma Synthesis and Processing of Materials

Edited by
Kamleshwar Upadhya

Plasma Synthesis and Processing of Materials

Proceeding of a symposium sponsored by
the Structural Materials Division of TMS,
held during the 1993 TMS Annual Meeting,
Denver, Colorado
February 22–25, 1993

Edited by
Kamleshwar Upadhya

A Publication of

A Publication of The Minerals, Metals & Materials Society
420 Commonwealth Drive
Warrendale, Pennsylvania 15086
(412) 776-9024

Printed in the United States of America
Library of Congress Catalog Number 93-077167
ISBN Number 0-87339-213-2

If you are interested in purchasing a copy of this book, or if you would like to receive the latest TMS publications catalog, please telephone 1-800-759-4867.

PREFACE

In the coming century, the single most important technical issue that will attract a great deal of attention will be the "Synthesis and Processing of Materials." The emphasis will be placed on the development of new materials by using futuristic technology as well as new processing routes which will conserve energy, be more productive, and be friendly to the environment. Plasma technology undoubtedly will play an important role in achieving these goals in the field of synthesis of nonequilibrium-compounds and processing of materials. Future applications of plasmas will be limited only by mankind's imagination.

In the United States, plasma technology received well deserved attention for its further development by the NASA Space program in the 1960's. NASA scientists needed a low mass-high temperature heat source to simulate temperatures of 8000 C and higher which spacecraft encounter during re-entry to earth atmosphere. NASA's dramatic achievement of putting astronauts on the moon and returning them safely to the earth was a glaring testimony to the potential capability of plasma heating technology and its applications in our daily lives in the coming years. Since then, there have been numerous developments in plasma devices and plasma assisted technologies which are used widely in engineering disciplines ranging from semiconductor industries, chemical, and materials processing fields.

This conference attempted to bring to the engineers and scientists engaged in plasma research the latest developments in this rapidly changing technology and to provide a forum for interaction and exchange of ideas and for sharing common problems and solutions in developing and refining the novel techniques of plasma processing of advanced materials. This goal was achieved by bringing together internationally recognized experts in plasma science and materials processing and highlighting the common goals in the areas of materials processing. This book contains invited and contributed papers on plasma fundamentals as well as plasma assisted processing of materials. These papers were presented by a broad group of researchers and scientists representing universities, federal and state research laboratories and industries. It is my hope that this book will be an excellent first reading for those engineers and scientists engaged in the plasma assisted processing of advanced materials.

The efforts of several people from TMS, in particular, Ms Marlene Karl and Ms Janet Urbis deserve a special thanks. I also wish to thank all authors, presenters, and participants who took part in the conference and by their contributions and ideas made the conference an overwhelming success. It was a most enjoyable experience and was made even more so, thanks to Professor P. R. Taylor, Department of Metallurgical and Mining Engineering, College of Mines, University of Idaho, Moscow, Idaho, 83843, who provided continuous support and help during the organization of this conference.

KAMLESHWAR UPADHYA
OLAC Phillips Laboratory/UDRI
Edwards AFB CA 93524-7680

Dedicated affectionately
to
"Swati" and "Indu"

Table of Contents

Section I

Section II

Section III

Section IV

Section I

"Plasma Processing of Ceramic and Ceramic - Matrix Composites"

K. Upadhya

UDRI/Phillips Laboratory
Bldg. 8424
Edwards AFB, CA 93523-5000

ABSTRACT

Due to the recent availability of plasma generating devices and their reliable performance, the applications of plasma technology in the materials processing is rapidly increasing. In this paper, therefore, the application of plasma technology in the materials processing has been critically examined. Also, the experimental results of the author's study on densification of ceramic materials in the plasma environment have been presented and discussed.

Aluminum nitride, silicon carbide, alumina and partially stabilized ZrO_2 -- with and without Al_2O_3 -- have been sintered in a radio frequency induction coupled plasma system. Translation speed of the specimens through the plasma, power level, gaseous dopants, sintering additives and pressure into the reaction tube were investigated as the experimental variables to elucidate the sintering mechanism(s). AlN, Al_2O_3 and partially stabilized Y_2O_3 + ZrO_2 rods have been sintered with final densities greater than 99.7% at a very high densification rate. The grain sizes of the sintered specimens were smaller by a factor or two than the same materials sintered in a non-plasma environment. Specimen temperatures were found to be a function of plasma-supporting gas pressure, gas composition, applied power and translation speed of the specimens.

Plasma Synthesis and Processing of Materials
Edited by Kamleshwar Upadhya
The Minerals, Metals & Materials Society ©1993

INTRODUCTION

Plasma processing of materials makes use of high energy content of the 'partially ionized gas' commonly referred to as the fourth state of matter. It has been estimated that 99% of the matter in the universe is in the plasma state. This may be due to the fact that stellar interiors and atmospheres gaseous nebulae and much of the interstellar hydrogen are in a plasma state. Plasma is also encountered in our daily life such as in a lightning bolt, the conducting gas inside a fluorescent tube, and inside a neon sign.

Plasma consists of charged, excited and neutral particles, resulting from the partial ionization of atoms or molecules of a gas and therefore, is an electrical conductor. However, any ionized gas cannot be called a plasma; as there is always a small degree of ionization in any gas. Therefore, a plasma can be defined more precisely as 'a quasineutral gas of charged and neutral particles which exhibits collective behavior'. Plasma provides a convenient source of in situ energetic ions and activated atoms which are now widely employed in densification and synthesis of ceramics, deposition and etching of materials.

The most important ionization processes in a gas are ionization by electron collision and by absorption of radiation, also called photoionization. Other process for ionization are; thermal ionization, cations collision, laser induced and collision with excited atoms and molecules. In collision processes if the collision is elastic, no excitation or ionization is produced. It is only inelastic collision that produces excitation and ionization i.e. a process in which part of the kinetic energy of the system before impact is converted into potential energy of one of the particles. The excitation and ionization energies of various gases is given in table I.

Generally plasma is classified in two groups:

(i) Thermal or equilibrium plasma. In this case the temperature of electrons and ions or atoms in the plasma is equal and the ratio of field strength (E) to pressure (P) is small. High intensity arcs, shock waves, nuclear fusion reactor and high pressure RF discharge fall into this group.

(ii) Non-thermal or non-equilibrium plasma is characterized by electron temperature much higher than the atoms or ions and a high value of E/P ratio. Glow discharge plasma, low intensity arc plasma, low pressure RF and microwave plasma fall into this group. The important parameters of both types of plasma are listed in Table II. In the film deposition technique usually glow discharge or low temperature plasma is utilized. In this type of plasma the degree of ionization is typically only 10^{-4}, so that the gas mostly consists of neutral or excited species. In glow discharge plasma electron temperature, Te, can vary widely and can be very high. for example, in Hg vapor rectifier, Te is $\simeq$ 15,000 ^{o}K and in neon tubes Te $\simeq$ 25,000 ^{o}K. In the plasma assisted depositon of materials, it is this extremely high electron energy which enables many reactions to occur at a much lower temperature compared to conventional PVD or CVD processes. It must be mentioned here that the extremely high temperature of electrons does not necessarily translate into a lot of

Table I. Some excitation and ionization energies of atomic and molecular forms of some gases

GAS		Excitation Energy (eV)	Ionization Energy (eV)
Oxygen	O_2	7.9	12.5
	O	1.97, 9.15	13.61
Nitrogen	N_2	6.3	15.6
	N	2.38, 10.33	14.54
Hydrogen	H_2	7.0	15.4
	H	10.16	13.54
Mercury	Hg	4.89	10.43
	Hg_2		9.6
Water	H_2O	7.6	12.59
Sodium Oxide	NO	5.4	9.5
	NO_2		11.0

Table II. Comparison of parameters of low-pressure and high-temperature plasmas

Parameters	Low-Pressure Plasma	High Temperature Plasma
Pressure	0.01 - 10 torr	760 torr
Discharge	0.001 - 1 A/cm^2	10 - 10^5 A/cm^2
current density		
Neutral Gas	77 - 1,300°K	10,000 - 40,000°K
Temperature T_n		
Electron	10^5-2x10^4°K	
Temperature T_e		
Vibrational	$\leq$5,000°K	
Temperaure T_v		
Rotational	$T_n \leq T_r, \leq T_v$	
Temperature T_r		
Degree of	10^{-8} - 0.8	~1
	Ionization	

heat. For example, in neon tube, Te $\simeq$ 25,000 ^oK, but the electron density is very small and therefore, total heat transfer to the wall by electron striking it at their thermal velocity is very small. Many laboratory plasmas have electron temperature of 100eV but density of 10^{12}-10^{13} per cm^3.

In the interest of readers unfamiliar with plasma a more detailed description of the plasma seems appropriate here. Consider an electric discharge plasma consisting of electrons ($Me=9.1x10^{-31}kg$), ions, excited atoms and neutral particles of plasma supporting gas ($Ma=6.7x10^{-26}kg$). It is well known that energy (E) associated with particles with zero velocity when placed in an electric filed (e) after time t, can be given by the equation:

$$E=q^2e^2t/M$$

where M= mass of particles .

Thus it can be seen from the above equation that the particles with lower mass will have higher energy associated with them. In other words, in electric discharge plasma electrons can gain higher energy much faster than ions or excited atoms. The energy from electrons is transferred to atoms by the collision process. At higher pressure, where particle density (N_d) is high, and the mean free path is short the collisions between electrons and heavy particles will be frequent and therefore, energy obtained by the electrons from the electric field is rapidly transferred to the heavy particles i.e. ions, excited atoms and neutrals. Thus, the temperature of electrons and heavy particles will be equal at high pressure or in the high particle density plasma and this type of plasma is termed 'equilibrium plasma. The equilibrium plasma such as nuclear fusion plasma and arc plasma have temperature in the order of 10^7 ^oK and 10^4 ^oK respectively.

As the density of particle decreases (N_d), (lower pressure) the mean free path increases and the number of collisions per unit time between electrons and heavy particles decreases, thus, reducing the energy transfer from electrons to the heavy particles. This results in the much higher electron temperature than gaseous atoms or ions and this type of plasma is termed 'non-equilibrium' or low-temperature plasma. Figure 1 illustrates the variations in the temperature of electrons and heavy particles with respect to pressure.

Currently, both types of plasmas i.e. low power plasma (<100kW) and high power plasma (>100kW) have found numerous industrial applications. The most important commercial application of plasma has been in the spray deposition of materials for protective coatings and more recently in manufacturing of near net shape components. Several others applications include production of submicron ceramic powders, heating, melting, and refining of metals, recovery of precious metals (Pt) from automobile catalytic converter (Texasgulf) and agglomeration of titanium scrap. The important parameters in application of plasma in the materials processing is the justification of the high cost of electrical energy which can be offset only by the high priced products. This may be one of the reason that plasma is widely used in semiconductor industry for etching and thin film deposition and to a limited extent in the production of submicron high purity carbides and nitrides powders. Another potential area in which plasma could be successfully utilized is in the processing of C-C and ceramic matrix composite materials. Latcher et al [1] have shown that C-C composite densified by D. C. arc plasma CVI yielded higher rate of densification at a much lower processing temperature in comparison to the conventional CVI process. This author studied densification of Y_2O_3+ZrO_2+MgO ceramic matrix composite in an induction coupled plasma [2] and found ICP technique extremely useful and beneficial in comparison to non-plasma conventional densification processes. Induction coupled plasma technique produced much higher densification rate coupled with an order of magnitude smaller grain size for the same materials densified in a non-plasma environment.

EXPERIMENTAL PROCEDURES

In the present investigation Al_2O_3 was selected as a model system because it has been well studied and widely used material. One would expect that the results obtained with Alumina will be applicable to a number of other ceramic materials. The compositions of Al_2O_3 and ZrO_2 powders are presented in Tables III and IV.

Alum-derived Alumina with specific surface area of $30m^2/g$ was used to from Al_2O_3 rods of 5mm diameter and 100mm long. 3 wt% polyvinyl butryal binder was added in aceton solution. 0.25% MgO was added via an aqueous aceton solution of $MgNO_3$. The Al_2O_3 rods were cold isotatically pressed at 300 MPa and presintered at 700°C for 1 hour. The increased toughness and high strength of Y_2O_3 partially stabilized zirconica has gained its reputation as a new structural material. These features are attributed to the martensitic transformation between the tetragonal and the monoclinic phases in the civinity of a rack tip which relives the fracture stress and consequently increases the fracture toughness. Lang [3] has reported that the application of hot isostatic pressing further improved the fracture strength of the Y_2O_3 PSZ + Al_2O_3 materials. However, he pointed out that the defects introduced by the fabrication process were responsible for the decrease in the fracture strength of both sintered and hot isostatically hot-pressed ceramic materials.

In this study, Y_2O_3 + PSZ with and without Al_2O_3 was employed to evaluate the effects of plasma sintering technique on the fracture toughness as well as other mechanical properties of this ceramic material. The partially stabilized 4.2 wt% Y_2O_3 + ZrO_2 + 20 wt% Al_2O_3 powder was cold isostatically pressed without addition of any binder at 300 MPa to obtain the rods of 10mm dia and 100mm long. These rods were also, presintered at 700°C for 1 hour.

A schematic of the Induction Coupled Plasma (ICP) sintering apparatus system is shown in Figure 1. The plasma was ignited within the quartz reaction tube by evacuating the tube to 200 millitorr and refilling the tube with Argon gas. Once the plasma was ignited the gas flow rate was increased to achieve a desired pressure in the reaction tube and maintained. The specimens were then sintered at different translation speed and power levels. The sintered specimens were examined for various physical and mechanical properties and their microstructure was studied by SEM.

RESULTS AND DISCUSSION

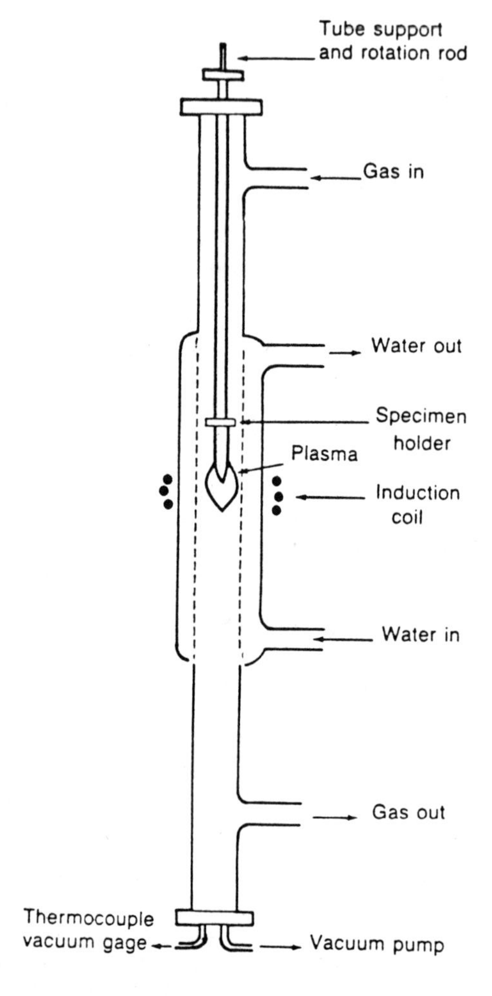

Fig. 1. Schematic of the ICP sintering apparatus system.

Table III Experimental Conditions

Al_2O_3 particle size	0.3mm
Specific Surface	$16m^2/g$
Power level	4 - 6 KW
Rods translation velocity	1 - 6 cm/min

Table IV Composition of Y_2O_3 stabilized $ZrO_2 + Al_2O_3$ (wt.%)

ZrO	75.2
Y_2O_3	4.2
Al_2O_3	20
SiO_2	0.01
Fe_2O_3	0.005
Na_2O	0.009
Ig Loss	0.6
Particle Size	0.3αμm
Specific Surface Area	30 m2/g
Crystal Phase	ZrO_2 (Tetragonal) + α - Al_2O_3

Composition of Y_2O_3 $PSZrO_2$ (wt.%)

ZrO_2	94.0
Y_2O_3	5.3
Al_2O_3	0.1
SiO_2	0.01
Fe_2O_3	0.002
Na_2O	0.007
Ig Loss	0.6
Specifice Surface Area	$31 m^2/g$

It is well-known that both impurities and stoichiometry departures can provide enhanced sintering of many ceramic materials [4-7]. Similarly, the sintering atmosphere can have a profound influence on the rate of sintering as well [8,9] as on the sintering mechanism(s). Also, the chemical composition of a given ceramic powder, particle size distribution and particle shape, all of these can affect the rate of sintering. Furthermore, any phase transformation difference in the diffusion rates in these phases, grain size and grain growth may also affect the rate of sintering of ceramic materials in a plasma environment. Before going in to detailed discussions of the experimental results, it is imperative that a brief outline of the sintering mechanism(s) be given here.

The sintering mechanisms can be divided into two groups. The one, in which coarsening is produced in the material without any densification and another that produces densification along with coarsening as well. In the first group belongs surface diffusion, vapor transport and lattice diffusion form the particle surfaces to the neck surface. The latter includes grain boundary diffusion and lattice diffusion from the grain boundary to the neck surface between particles, especially during the initial stage of sintering. These phenomenon have been shown schematically in figure 2. Although accurate surface diffusion data generally are not available in the literature, there is a general thought that the activation energy for surface diffusion, especially at low temperatures, is less than that for grain boundary diffusion which in turn is less than that for the lattice diffusion [10]. Thus, coarsening of ceramic materials will be significant especially at the lower temperature range at which atomic movement is predominant.

Coarsening has deleterious effects on the densification during the sintering of ceramic materials. Firstly, as the grain becomes coarser, the local surface curvature decreases thus decreasing the driving force for the sintering reaction. Also, the average diffusion distance for mass flow from grain boundaries to pores increases, thus further reducing the rate of densification.

In the present investigation a high percentage of densification at a very high rate as well as very fine grain sizes have been obtained for both Al_2O_3 and ZrO_2 plasma sintered specimens (Figure 3). The possible explanations for a high level of densification and fine grain sizes observed for the materials sintered in plasma environment could be due to the suppression of surface diffusion by extremely fast translation of ceramic materials through the high temperature zone (plasma flame) and thus preventing the coarsening of ceramic materials. During conventional sintering, on the contrary, a significant time is spent in the temperature range where the coarsening effect of surface diffusion outweigh the densification produced by the grain boundary and lattice diffusion.

As seen in Fig. 4 MgO doped Al_2O_3 showed less porosity than undoped specimen and the grain size is smaller by one magnitude than undoped specimens. Figure 5 illustrates the effect of MgO doping on percentage radial shrinkage of Al_2O_3 specimens. It is clearly seen that although the shrinkage is independent of the translation speed of the specimen through the plasma flame and also on the MgO doping of Al_2O_3 however, the translation speed into the plasma did affect the final density of the specimen. As the TS increased up to 3

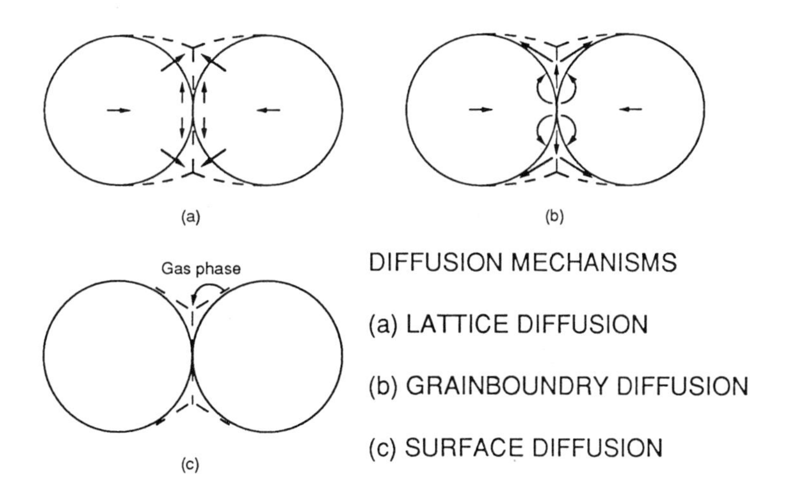

Figure 2 - Diffusion mechanisms prevailing during the sintering process.

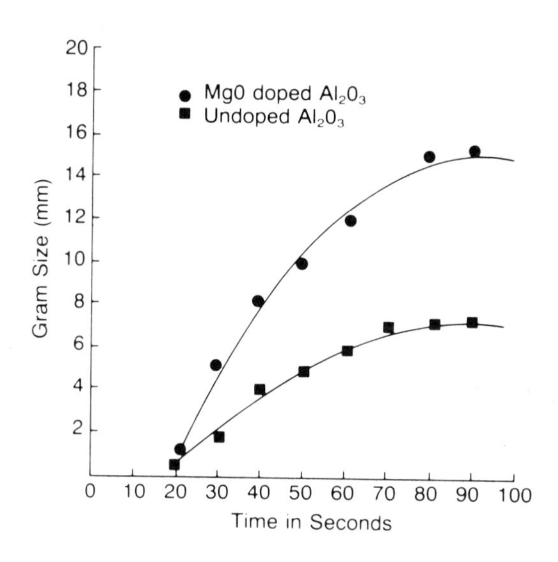

Figure 3 - Effect of MgO doping on the grain size of sintered specimens.

Figure 4 - Effect of MgO doping on the porosity of sintered Al_2O_3 specimens.

cm/min. the density increased but further increase of TS led to the decrease in the final sintered density of the specimens. This is illustrated in Fig. 6.

Figure 7 shows the SEM micrograph of a specimen of partially stabilized ZRO_2 specimen (4.2 wt% Y_2O_3) containing 20 wt% Al_2O_3. The grain size for this specimen were found to be less than 1 μm. This is much finer grains than for the identical material sintered by convention technique in non-plasma environment which grain size are 1-2 μm. Figure 6 illustrates the percentage linear shrinkage rate against time. It can be seen that the rate first increased reached a peak and subsequently decreased. This is because during the first stage, only densification occurs without any significant grain coarsening. In the second stage, the extent of coarsening is greater than the degree of densification. Furthermore, as the grains become coarser, the local grain boundary curvature decreases thus increasing the distance of lattice diffusion and thus, lowering the driving force for the sintering reaction and therefore, the rate of shrinkage in the specimens. However, the final total percentage shrinkage for both with and without Al_2O_3 PS zirconia specimens were between 16-18 percent.

Figures 8 and 9 are the plots of sintered density against time, for the PS ZrO_2 specimens with and without 20 wt% Al_2O_3. As it is clearly observed that the ZrO_2 specimens in both cases attained very high sintered density and it was 97-98.5 of theoretical densities of the materials. It is worth mentioning here that extremely high densities value are obtained without any grain coarsening which is inherent part of the conventional furnace sintering into a non-plasma environment.

One very different characteristic of ZrO_2 sintering was the effect of translation speed of the specimen on the sintered density of the materials. In Al_2O_3 specimens the final density increased reached a peak at 3 cm/min. and then decreased at higher translation speed up to 6 cm/min. However in ZrO_2 specimens, as the translation speed increased the final sintered density also increased. This anomaly may be attributed to the difference in the specific surface area of these specimens materials, presumably Al_2O_3 had larger specific surface area than ZrO_2 specimens.

One common features of both materials was the spontaneous cool down of the specimen by 600-700^oC in the plasma if the specimens were suddenly stopped and held into plasma. This could be attributed to the plasma penetrating the specimens pores and in - situ generation of the heat into the pores. Also, the higher temperature at higher translation speed of the specimens could be explained in terms of exposure of higher surface area of the specimens before closure of the pores, thus permitting the attainment of a higher temperature. Also, Pfender et al [11] attempted to explain this 'cool down' phenomenon by modeling the heat transfer of the plasma sintering processes. Their finding certainly lends this author above mentioned explanation. Pfender et al concluded: (i) heat transfer from plasma to the specimen depended strongly on the surface area and the catalytic properties of the specimens. (ii) High heating rate alongwith low thermal conductivity of green samples resulted in the higher surface temperature of the specimens (iii) the maximum temperatures of the sintered specimen held stationary in to plasma were lower because the heat transfer coefficients for sintered specimens were lower than those of green samples. (iv) Higher zoning speed caused higher heating rates which increased the total heat flux to the

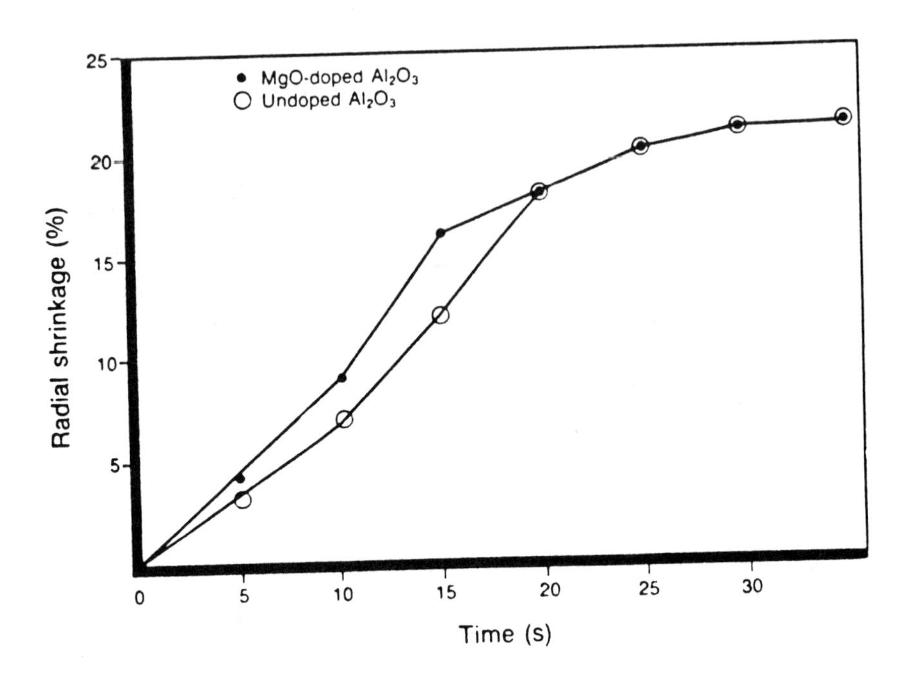

Figure 5 - Effect of MgO doping on the percentage radial shrinkage of Al_2O_3 specimens.

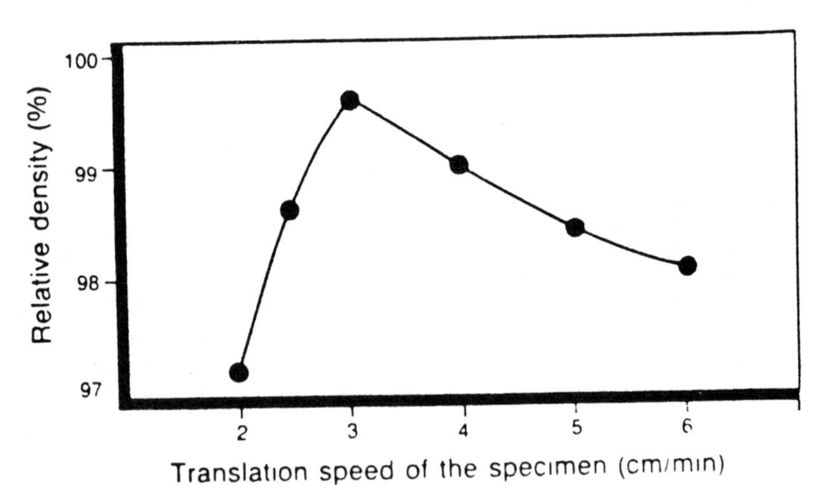

Figure 6 - Effect of translation speed of the specimens through the plasma on the final sintered density of $Y_2O_3 + ZrO_2O_3$ specimens.

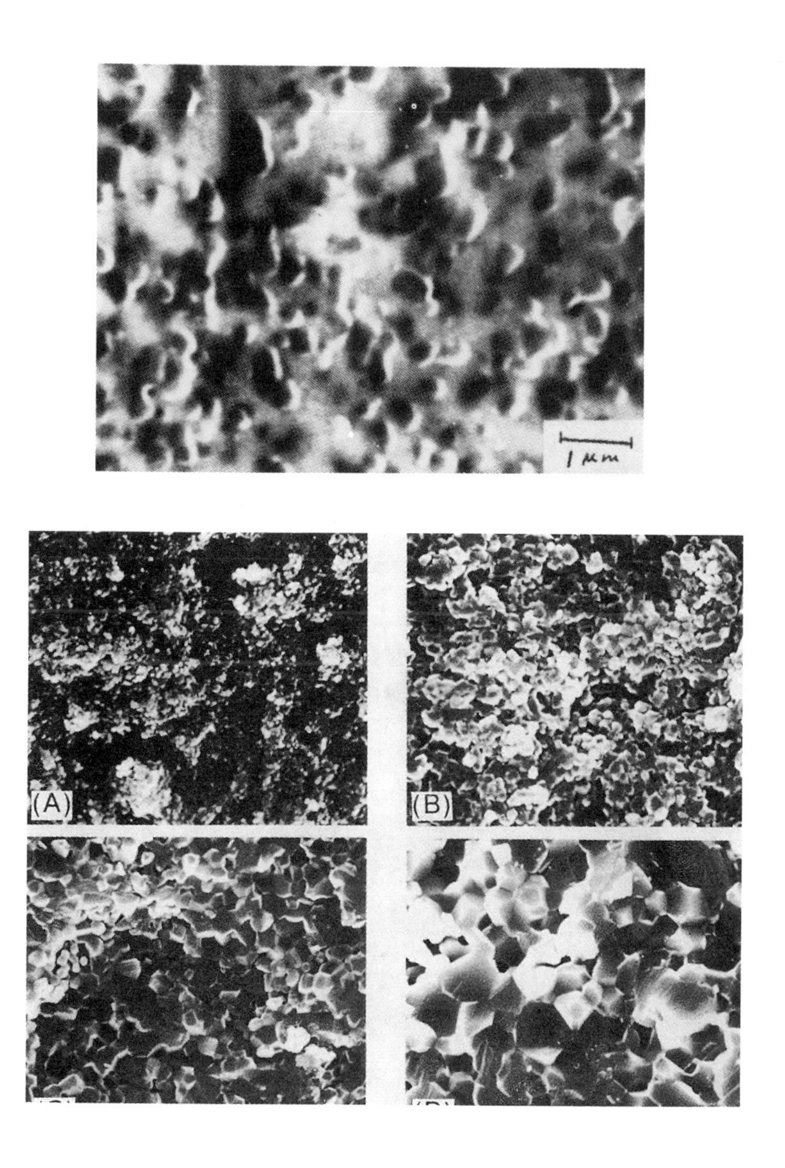

Figure 7 - (a) microstructure of $Y_2O_3 + ZrO_2$ specimens (b) microstructure evolutions during sintering of Al_2O_3 specimens with 15 seconds intervals.

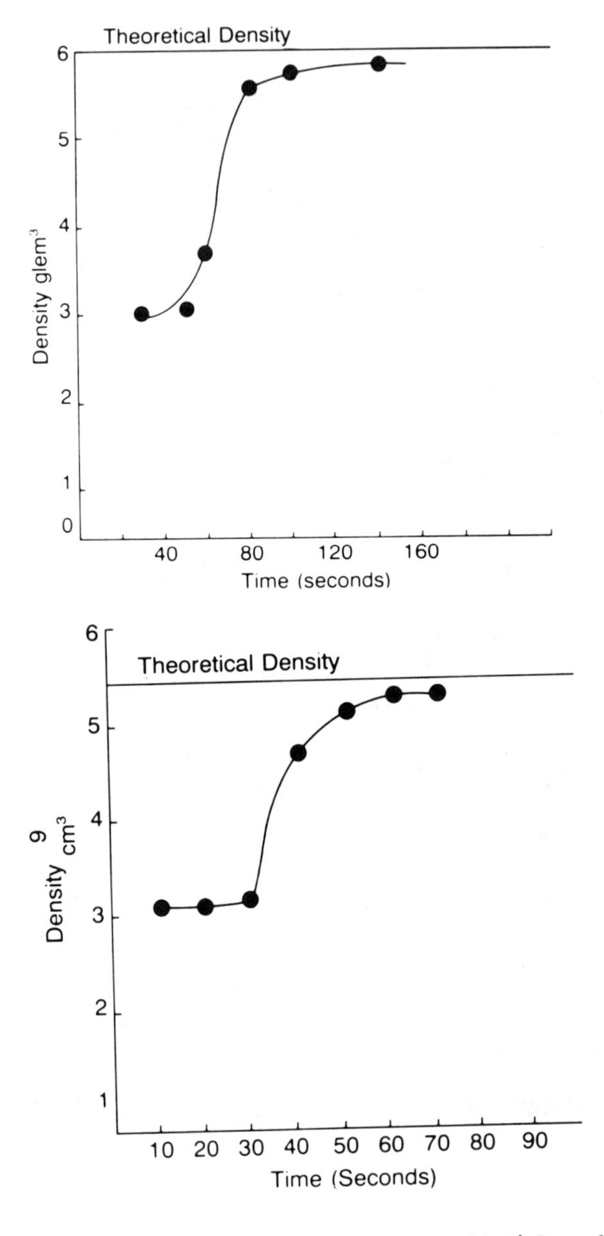

Figure 8&9 - Sintered densities of Y_2O_3 +ZrO_2 specimens with Al_2O_3 and without Al_2O_3 additives.

specimens. It must be pointed out that the above findings may be only partial reasons for the abnormal plasma effects observed and a complete picture still remains unclear.

NON-OXIDE CERAMICS MATERIALS

As mentioned earlier, most of the research studies have been performed on the Al_2O_3 or ZrO_2 based ceramic or ceramic-matrix composites in the plasma environment. Very few studies have been reported on the non-oxide ceramic materials. Amongst non-oxides group only SiC, AlN and Si_3N_4 have been studied in a plasma system. Kijima [12-13] reported densification of Boron and Carbon doped SiC of very fine starting powder with a mean particle size of 0.3μm in an Ar plasma at a pressure range of 1.3 Pa to 130 Pa and apparently obtained very high densities in less than 10 minutes. There were no details on the translation speed of the specimen through plasma, initial binder material or pressure level applied for powder compaction. However, he mentioned that the final sintered density was dependent on the argon flow rates and the plasma supportive gas pressure. For a fixed pressure,and increasing gas flow rate the specimen density increased and with further increase in gas flow rate it decreased. Also, at a constant gas flow rate, the densities of sintered specimens decreased with increasing pressure after establishing a maximum at 48kPa. However, no explanation was offered by Kijima et al for the above mentioned phenomenon. This author believes that an increase in the gas flow rate and the pressure both of these plasma parameters induce the instability in the plasma for a given power level and causing the observed decrease in the sintered densities of the specimens. Kong [14] studied SiC sintering in the Induction Coupled Plasma with boron and carbon as additives in an $Ar/He/H_2/N_2$ mixed plasma at 100kPa. After an hour sintering period he obtained 90% density of the specimen but did not report on the grain size of the sintered specimens. This author sintered SiC with boron and carbon in an ICP in an $Ar/He/N_2$ plasma at 100kPa and obtained 92% of theoretical density after 35 minutes of sintering period with average grain size of 1.0-1.5 μm.

AlN due to its relatively higher thermal conductivity has attracted tremendous attractions in semi conductor industries. Therefore, many successful attempts has been reported of plasma sintering of AlN achieving full density and much finer grain size in comparison to the specimen sintered in a conventional non-plasma environment. Kijima [15] sintered AlN with Y_2O_3 as sintering additive in an Ar IC plasma at 0.14kPa and reported achieving full density in only 60 seconds of sintering period. Knittle [16] sintered AlN with and without Y_2O_3 in a nitrogen plasma at 13kPa. The sintered density of 95% and 81% were reported for doped and undoped specimens. However, when Knittle [17] sintered Y_2O_3+AlN in a microwave plasma for 15 minutes he obtained full density with grain size of 2μm compared to a grain size of 8μm for the similar specimen sintered in a conventional N_2 non-plasma environment.

CONCLUDING REMARKS

Plasma processing of ceramics and ceramic matrix-composites is a relatively new technology with great potential. However, this technology is still in its infancy because a large number of observed phenomenon remain unexplained and not very well understood. Therefore, a lot of research efforts need to be directed to understand the fundamentals of the nucleation and growth phenomenon of materials when processed in to the plasma environment where extreme rapid heating and chemical reactions are occurring. Also, plasma diagnostics studies have to be undertaken to isolate the synergistic effects of the most important parameters of any plasma i.e. electron density, energy and distribution function. In conclusion, plasma processing of ceramics has proven itself an important tool as it produces ceramic materials with much higher densities coupled with finer grain size in a much shorter period of processing in comparison to conventional established processing techniques.

REFERENCES

1. Lachter, A., Trinquecoste, M., and Delhaes, P., "Fabrication of Carbon-Carbon Composites by D. C. Plasma Enhanced CVD of Carbon" Carbon, vol 23, No.1 (1985) P. 111-116.

2. Upadhya, K. "An Innovative Technique for Plasma Processing of Ceramic and Composite Materials" Ceramic Bulletin, vol 67, No. 10 (1988) P. 1691.

3. Lange, F. F.., "Processing Related Fracture Origins": I, Observations in Sintered and Isostatically Hot Pressed Al_2O_3 +ZrO_2 Composites" J. Amer. Cer. Soc., 66 (6) (1983) P. 396.

4. J. F. Shackelford, P.S. Nicholson, and W. W. Smeltzer: J. Am. Ceram. Soc., 1974, 57, 235.

5. S. B. Boskovic and M. M. Ristic: Sov. Powder Metall. Met. Ceram., 1972, 11, 755-759.

6. S. B. Boskovic and B. M. Zivanovic: J. Mater. Sci., 1974, 9, 117-120.

7. S. B. Boskovic, M. C. Gasic, and M. M. Ristic: in "Modern Developments in Powder Metallurgy," vol 4, (ed. H. H. Hausner), 357-36, 1971, New York, Plenum Press.

8. J. P. Roberts, J. Hutchings, and C. Wheeler: Trans. Br. Ceram. Soc., 1956, 55, 75-79.

9. K. Nii: Z. Metallkd., 1970, 61, 935-941.

10. D. L. Johnson, "Ultra Rapid Sintering" in "Sintering and Hetrogenious Catalysis" Ed. G. C. Kuczyaski, A. E. Miller and G. A. Sargent, Plelnum Pres 1984.

11. E. Pfender and Y. C. Lee, MRS Proc., vol 30, P. 141 (1984).

"Plasma Assisted Deposition and Synthesis of Novel Materials"

K. Upadhya
UDRI/Phillips Laboratory
Bldg. 8424
Edwards, CA 93524

ABSTRACT

Due to the recent availability of plasma generating devices and their reliable performance, the application of plasma in the thin film deposition technology is rapidly increasing. Plasma assisted deposition techniques have witnessed a surge in the research activities for processing thin films of super-conducting materials: Diamond or diamond-like carbon (DLC), Cubic boron nitride and B-SiC. Recent evolution's and developments in the plasma assisted deposition technologies have been received. The current status is outlined in terms of process parameters, their control, flexibility in processing the types of films and product quality. The advantages and limitations of plasma assisted deposition processes (PADP) in terms of plasma interactions at the source, during materials transport and at the substrate during film deposition has been critically examined. Also, synthesis of novel materials have been broadly outlined and discussed.

Plasma Synthesis and Processing of Materials
Edited by Kamleshwar Upadhya
The Minerals, Metals & Materials Society ©1993

Introduction

In the era of current high technology, the tendency to use a composite material, where the surface properties are intentionally different from those of the bulk material is ever increasing. As a result, the last two decades have witnessed an amazing growth in methods to produce unique surface properties while preserving the bulk properties of the core material. These techniques have made increasing utilization of plasma, Laser, Ion and Electron Beams, and chemical and physical vapor deposition processes. The materials with modified surface are used in a wide variety of spectrums ranging from electronic devices, semi-conductors, solar cells, thermal insulation tribology and in decorative coatings, to name a few. Materials used to modify the surfaces by coatings, cladding and implantation are gaseous species, metals, alloys, refractory compounds, intermetallics and polymers. The thickness of implanted or coated surface may vary from a few °A to several mils.

These modern techniques can be broadly grouped into two categories, one which uses plasma, and others utilizing laser as the processing tool. However, in this paper, only plasma assisted deposition techniques will be discussed. These techniques essentially are variants of chemical and physical vapor deposition processes which rely on vapor transport of materials to construct new surfaces. The mean energy E, of a gas atom with respect to temperature is given by

$$E = 3/2\ \kappa T$$

where κ = Boltzman constant

T = Temperature in °κ

Thus, vapor transport of materials for constructing new surfaces by simply increasing in temperature is not promising, as it will raise the substrate temperature correspondingly. Therefore, plasma assisted ionization of the vapor atoms of the material to be coated with will provide the alternative suitable method for coatings. The ionized vapor species have greater reactivity and also, they can be accelerated in the electric field and the combination of these two properties will provide a coating with better physical and mechanical properties.

Therefore, the main theme for developing a powerful efficient coating process is clear: Ionize the coating species and transport them to the surface to be modified in an electric field. When ionization of species are used for the surface modification, then two different methods can be adopted for coating the substrate material. In the first option, a small ion flux with high mean energy per ion is used, while in the second option, a high ion flux with sufficient mean energy per ion to ensure a good adhesion of the coating to the substrate is used. The first option is called "Ion Implantation", while the other is termed "Plasma Assisted Coating" processes, i.e. Ion Plating, PACVD and Ion Beam Mixing, etc. It is worth noting that in both instances, there is an upper limit to the energy flux which can be deposited on the surface to be modified if the substrate properties are to be kept unaltered.

Ion Implantation is the process whereby one or more ionic species are introduced into the surface of the material by using a low flux, but high energy, ion beam. This results in modified physical and chemical properties of a very thin near surface region. Since Ion

Implantation is a non-equilibrium process, the thermodynamic solubility limits of the system may be exceeded with or without subsequent precipitation. Theoretically, it is possible to implant any kind of ion into a substrate material.

On the other hand, Ion Plating is a deposition process in which the substrate surface is coated with a film deposited from a high ion flux with sufficient energy of the beam obtained from an activated source and a preinduced ionized plasma. Table I lists the established and relatively newer surface modification processes. Table II shows main characteristics of various deposition techniques for the compound films.

The parameters which govern the performance of the coatings fall into two groups, namely the volume properties which relate to the contacting bodies as a whole, and the surface properties which determine the contacting interface of these bodies. In the first group, the important properties are yield strength and hardness followed by Young's modulus, shear modulus and stored elastic energy. The other properties of interest may be the ratio of fracture stress in tension to yield stress in compression. The thermal properties must be also taken into account. In the second group falls the chemical reactivity and the adsorption properties of the surface of the material.

Plasma Assisted Deposition Technologies:
Theoretical Background

The uniqueness of the Plasma Assisted Deposition Processes lies in their ability to synthesize films at relatively low substrate temperatures (300-500°C). These processes offer the possibility of varying film properties over a wide range by suitably controlling the plasma parameters, i.e., electron density, energy and distribution function. Therefore, Plasma Assisted Deposition Processes have witnessed the phenomenal growth in the industrial applications in depositing films such as dielectric, metallic, semiconductor, micro-electronics, optics and optoelectronics, hard carbides and nitrides, sulfide films for solid lubricants, especially in the space applications and in solid state batteries.

Essentially, there are three steps in the formation of a coating film on the substrate:

1. Synthesis or creation of the depositing species
2. Transport of these species from the source to the substrate
3. Nucleation and growth of the film onto the substrate

This model is shown schematically in Figure 2.

Table I

Methods of Coating Deposition

Atomistic	Particulate	Bulk Coatings	Surface Modification Techniques
Electrolytic	Thermal Spraying	Wetting Process	Chemical Conversion
Electroplating	Plasma spraying	Painting	Electrolytic
Electroless plating	Detonation gun	Dip coating	Anodization (oxide)
Fused salt electrolysis	Flame spraying		Fused salt technique
Chemical displacement			
Vacuum Environment	Fusion Coating	Electrostatic	Chemical Liquid
Vacuum evaporation	Thick film ink	Spraying	CVD
Ion Beam deposition	Enameling	Printing	
Molecular beam epitaxy	Electrophoretic	Spin Coating	Thermal
			Plasma
Plasma Environment	Impact Plating	Cladding	Mechanical Treatments
Sputter coating			
Activated reactive evap.	Explosive Bonding	Shot Peening	
Plasma polymerization	Roll Bonding	Thermal	
Ion plating			
Chemical Vapor Environment	Overlaying		
Chemical Vapor Deposition	Weld Overlay		
Reduction			Surface Enrichment
Decomposition			
Plasma enhanced			Diffusion from bulk
Spray Pyrolysis			Sputtering
Liquid Phase Epitaxy			Ion Implantation

Table II

Typical Characteristics of Established Deposition Processes

Characteristic	Chem.Vapor. Dep.	Electro-Deposition	Thermal Spraying
Mechanism of species production	Chemical reaction	Deposition from solution	From flames or plasma
Deposition rate	Moderate (200-2,500 Å/min)	Low to high	Very high
Species deposited	Atoms	Ions	Droplets
Throwing power for:			
Complex shaped objects	Good	Good	No
Into small, blind holes	Limited	Limited	Very Limited
Metal deposition	Yes	Yes, limited	Yes
Alloy deposition	Yes	Quite limited	Yes
Refractory Comp. deposition	Yes	Limited	Yes
Energy of species deposited	Can be high with	Can be high	Can be high
		plasma-assisted CVD	
Bombardment of substrate/deposit by inert gas ions	Possible	No	Yes
Growth interface (by external means)	Yes (by rubbing)	Yes	No

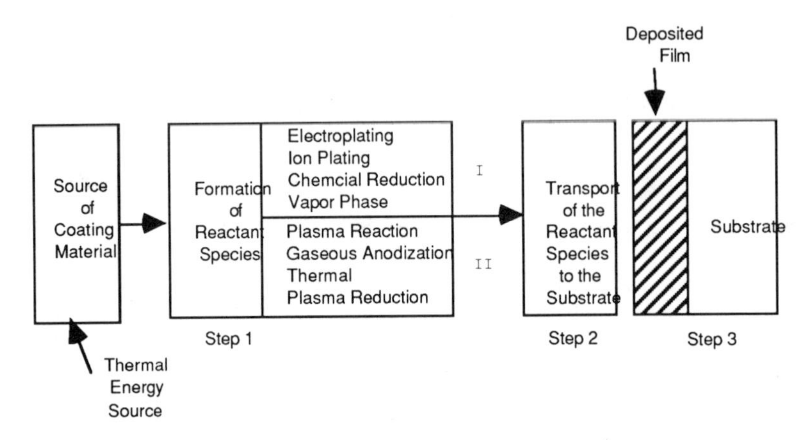

Figure 1. Schematic Representation of Three Steps Involved in Coating Fabrication

These steps can operate independently or interdependently, depending on the particular process. However, it is preferable to have a process where these steps operate independently, thereby allowing greater flexibility and control and providing more degree of freedom over a process where these steps operate interdependently.

The Role of the Plasma

Increasingly, plasmas have been used as an important tool in materials processing during the last two decades. The main attractiveness for plasma utilization is its ability to provide an "in-situ" source of ions, activated atoms and energized molecules at a relatively low temperature, which enhances the various physical and chemical processes and thus producing a deposited film with more desirable properties. The major functions of the plasma in the film deposition technique are: (i) Generation of the vapor species, i.e., sputtering deposition technique, (ii) Activates and enhances the reactions for deposition of metal or compound films, and (iii) modify the structure/morphology of the films. As mentioned earlier, plasma have advantageous as well as deleterious effects on the process parameters depending upon the deposition techniques; for example, in addition to enhancing the reaction kinetics in the reactive sputtering deposition, it may also produce "target poisoning", i.e., over a critical flow rate of the gas fc, the deposition rate decreases by a factor of 3 or more. The critical flow rate is a function of power to the target, pressure into the chamber, and target size.

The relationship between the film properties and gas injection rate is non-linear and it is believed to be due to a complex phenomena involving dependence of the sticking coefficient on the growth rate, composition structure, and temperature of the growing film [1]. Hence, the selection of a plasma assisted process for a particular application is not a straightforward decision. Some of the important factors which should be included for consideration are as follows:

(1) The materials to be deposited.

(2) Deposition rate required.
(3) Maximum substrate temperature tolerable.
(4) Adhesion of the film to the substrate.
(5) Throwing power of the process
(6) Facility for substrate bombardment for the modification of structure/morphology of the depositing film.

The presence of the plasma in the source-substrate vicinity greatly affects the process parameters at each of the aforementioned three steps of film deposition. However, the effect of the plasma on each of these three steps will be different in terms of the types and concentration of the ionized species, activated atoms, energized molecules and radicals, which in turn will affect the reaction paths and the distributions and physical locations of "active reaction sites" on the surface of the substrate. Also, as noted by Deshpandey, et. al., [2] that ionization probability of atoms is maximum if the electron energy is in the range of 50 - 60 eV and decreases with further increase in energy. Therefore, it is preferable to have low energy electrons as is the case in ARE type processes. Bunshah [3] has shown that the advantages and limitations of various plasma assisted deposition techniques can be addressed in terms of the differences in the plasma interactions at the source, during transport, and at the substrate during film deposition in the various processes. Table III shows the comparisons between three most commonly used plasma assisted deposition techniques, i.e. Reactive Sputtering (RS), Activated Reactive Evaporation (ARE) and Plasma Assisted Chemical Vapor Deposition (PACVD) in terms of plasma/source - plasma/volume and plasma/substrate inter-actions. Also, the table reflects the advantages and limitations inherent to each of these processes.

(i) Plasma-Source Interactions

In sputtering deposition, for example, vapor species are generated by momentum transfer from positive ions bombarding the target and thus knocking out atoms from the target material. Therefore, the rate of vapor species generation will be dependent on the power input from the plasma to the target, i.e., the cathode voltage and current in the case of DC and RF sputtering. Thus, sputtering rate, in this case, is totally plasma dependent. On the contrary, in Ion Beam Deposition Technique, the sputtering rate is independent of plasma. In the ARE process which is developed and perfected by Bunshah, et.al., [4,5] metal atoms are produced by evaporation from the source which may be heated by a Thermionic electron beam, a plasma electron beam, resistance heating or arc heating. In this process plasma is created in the source-substrate space by injecting low energy electrons (20-200 eV). The source of the low energy electrons can be a thermionically heated cathode or the plasma sheath above the molten metal pool with an appropriately spaced anode biased to a low positive potential. RF can also be used to form plasma. Thus, it is clear that in ARE process, the vapor species are generated by thermal energy imparted to the source material. The rate of generation of vapor species varies directly with the vapor pressure of the source material, which in turn will be dependent on the surface temperature of the source material. The plasma has little effect on the rate of evaporation of source material. Also, numerous studies have been reported on specific metal-reactive gas combinations to elucidate the role of physical and chemical processes occurring during reactive sputtering [6-25]. The general consensus among all these studies is that plasma/target reactions may play an important role in an overall RS reaction mechanism(s).

Table III

A Comparison of Physical Vapor Deposition Processes

Characteristics	Evaporation	Ion Plating	Sputtering
Mechanism of species production	Thermal energy	Thermal energy	Momentum transfer
Deposition rate	Can be very high (up to 75 μm/min)	Can be very high (up to 25 μ/min)	Low expect for pure metals (Cu: 1 μm/min)
Species deposited	Atoms and ions	Atoms and ions	Atoms and ions
Throwing power for:			
Complex shaped objects	Poor: Line-of-sight coverage except for gas scattering	Good but non-uniform thickness distribution	Good, but non-uniform thickness distribution
Into small, blind holes	Poor	Poor	Poor
Metal deposition	Yes	Yes	Yes
Alloy deposition	Yes	Yes	Yes
Refractory Comp. deposition	Yes	Yes	Yes
Energy of deposited species	Low	Can be high	Can be high
Bombardment of substrate/deposit by inert gas ions	Not normally	Yes	Yes or no depending on geometry
Growth interface perturbation	Not normally	Yes	Yes
Substrate heating (by external means)	Yes, normally	Yes or No	Not generally

(ii) Role of Plasma During Material Transport

During the plasma-vapor species interactions, there are three important reactions occurring simultaneously. These are electron impact ionization, excitation of atoms and molecules, and dissociation.

These can be represented as follows:

$$e^- + A = A^* + e^- \qquad \text{(Excitation)} \qquad (1)$$

$$e^- + A = A^+ + 2e^- \qquad \text{(Electron Impact-Ionization)} \qquad (2)$$

$$e^- + A = B^+ + C + 2e^- \qquad \text{(Dissociation)} \qquad (3)$$

The rates of these reactions can be represented [26] as:

$$R = N_e\ K_i\ [A] \qquad (4)$$

where R = Rate of reaction
Ne = Electron concentration
K_i = Rate constant
$[A]$ = Concentration of A

Further, the rate constant for these reactions has been shown to be [27]

$$Ki = (2/me) \int E f(E)\ i\ (E)\ dE \qquad (5)$$

where Me = Electron mass

$f(E)$ = Electron distribution function
$i(E)$ = Collision cross section for a particular reaction

Thus, from equations (4) and (5), one can easily estimate the rate of formation of a particular species in the plasma. Thornton [28], in an excellent publication, has described in detail the analytical model and illustrated the principle mechanism of radical formation in a plasma. Numerous types of radicals metastable species as well as excited and ionized species may be generated in the plasma and it is shown in Table IV.

Plasma - Substrate Interactions

The substrates, when exposed to a plasma, will be bombarded by ions, electrons and energized atoms. As discussed by Holland [29], the nature and energy of these bombarding species will be dependent on the process parameters and geometrical aspect of the substrate within or outside plasma zone. As a result of substrate bombardment, numerous changes may take place, such as substrate heating, substrate surface cleaning, re-emission or sputtering of deposited material, gas incorporation in the depositing film, and

Table IV

Products and Activated Species
Expected in a Plasma Environment at 8000°K

Source	Plasma Gas Argon	Elements Carbon	Compound	
Products	Ar^+	C_2, C_3 *etc.* C^+, C^{2+} *etc.* S^+, S_2, S		FeO, Fe, Fe^+ O_2, O, O^+, O^- TiO_2, Ti_2O_3, Ti_3O_4, Ti, Ti^+, Ti^{2+} *etc.*

Table V
Synthesis Techniques for Novel Materials

NOVEL MATERIALS	Processing Techniques ARE	RS	PACVD	Others
Superconducting Materials such as Nb Ge, CuMO S	1000-1500Å/min			
Photovoltaic Materials A-SiH, CuInS , etc.	1500-2000Å/min	50-200Å/min		
Optoelectronic Materials Indium Tin Oxide, Zinc Oxide, etc.	500-1000Å/min	60-150Å/min		
Cubic Boron Nitride	1000-1500Å/min			1500Å/min[1]
Diamond, or DLC			1000Å/hour	10,000Å/hour[2]
I-C		300Å/min	200Å/min	2μm/hour

[1]ADRRP
[2]MPACVD

modification in the film structure/morphology. Thus, plasma-substrate interactions may have significant effect on the final properties of the growing film.
The main reason for the substrate bombardment in a plasma or glow discharge is due to the potential developed on the surface of the substrate with respect to the plasma. Due to the difference in the mobility of electrons and ions, a space charge region forms (C in Figure 2) adjacent to the surface in contact with the plasma from which one species (electrons) is largely excluded.

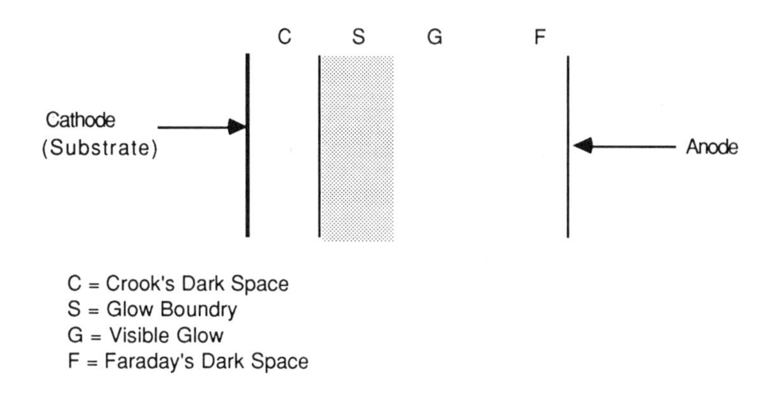

Figure 2. Schematic Illustrations of Various Zones Associated with a Plasma

The nature of this region 'C' (sheath) depends mainly on the current density passing across it. The main role of this sheath is to produce a potential barrier so that electrons are electrostatically deflected away from the substrate. The magnitude of this potential barrier will adjust itself in a manner to balance the electron flow to the substrate to be equal to that drawn out in the external circuit. Thus, it is true that any surface coming in contact with plasma will develop a potential which will be a little negative with respect to the plasma. Thus an electrically isolated surface in contact with the plasma develops a negative potential with respect to plasma so that the electron flux equals to the ion flux. Generally, this potential is referred to as the "Floating Potential". The significance of this analysis is that the substrate bombardment by species is dependent on this "Floating Potential", which itself is dependent on the electron energy and distribution functions. Thus, to exert control over the substrate bombardment and therefore, on the various substrate reactions initiated due to this, which modify the nucleation and growth phenomenon of the film, the electron energy and distribution function can be varied independently of other process parameters.

Plasma has found application in both major types of deposition techniques, i.e., Chemical Vapor Deposition (CVD) and Physical Vapor Deposition (PVD) processes. In the Plasma Assisted CVD (PACVD); R.F., microwave and photon (U.V., Laser) excitation have been used to generate plasma. However, in the PAPVD, D.c. R.F., triode, magnetron geometries and reactive sputtering have been used. Also, Bunshah, et.al [30,31] have developed a very useful deposition PAPVD technique, namely Activated Reactive Evaporation (ARE) process, in which metal atoms are produced from an evaporation source and may be heated by a Thermionic electron beam, a plasma electron beam, resistance heating, or arc heating. The gas employed is only reactive gas at a pressure less than 10^3 torr. The plasma is generated in the sourcc-substrate space by simply injecting low energy electrons (20-200 eV).

Synthesis of Novel Materials

Materials possessing unique combinations of physical, mechanical, electrical and optical properties are classified as "Novel Materials". A few examples of such materials are:

i) Diamond, which is electrically insulating but has extremely high thermal conductivity.
ii) Materials with high hardness value and high ductility such as microlaminate composite
iii) Materials with metastable structure [32]; such as B-SiC, Cubic-BN and synthetic diamond and similar materials.

It will be nearly impossible to discuss all the materials and coatings synthesized by using plasma techniques. However, "Novel Materials" synthesized by plasma assisted techniques are listed in Table V. A few of these will be discussed in detail here. The unique combination of optical, physical and electronic properties of Diamond has been the main driving force for the development of new and less expensive methods for diamond syntheses. Potential application for diamond include reliable infra-red window coatings, high temperature semiconductors, high energy microwave amplifiers, powerful pulsed lasers, wear coatings, and as a heat sink material. Diamond film is hard, stiff, and has very low coefficient of friction. It can transmit light from the far infra-red through ultra-violet. It is also one of the few known materials that is both an excellent thermal conductor but an electrical insulator. Aerospace manufacturers may be the first to use diamond thin film as a coating material on optical windows and domes. Diamond film will protect the fragile sensors and lasers from the outside hostile environments. These devices are used in heat seeking missiles, satellites and in night vision systems. Most of these devices operate in the FIR wave range, i.e., 7μm and upwards, which permit them to see through clouds and distinguish cooler objects from others.

The optical properties of window materials gets destroyed under stress. For example, Germanium distorts when heated by supersonic flight or high-powered lasers. Zinc Selenide flakes and spalls off on impact. Zinc Sulfide has better mechanical properties but a narrower FIR bandwidth. Diamond solves most of the above mentioned problems. It is almost transparent to FIR waves with 7μm and greater, and is hard enough to shed rain and ice at hypersonic speed. Also, a high refractive index with 2.4 makes it a suitable window material with integral lenses.

There are several methods, such as Plasma Assisted Chemical Vapor Deposition (PACVD) [33-39] Ion Beam Deposition [40,41] sputtering [42] and hot filament techniques [43] which have been employed for diamond film fabrication. Different types of Diamond for different applications can only be synthesized by different techniques of diamond crystallization. For example, diamond for high temperature semiconductor application can be prepared by creating alternating layered structure of dielectric and semiconducting single crystalline diamond films, with the thickness probably in the range of one μ inch. At the lower temperature, lower pressure diamond synthesis, the possibility of nucleation for both diamond and graphite crystal exist equally. Therefore, in the PACVD for the syntheses of diamond, the presence of super equilibrium concentration of atomic Hydrogen is critical and this is obtained by the dissociation of H_2 into the plasma. When the concentration of atomic Hydrogen is at least one order of magnitude higher than that for thermal equilibrium, nucleation of graphite is then suppressed. A plasma rich in atomic hydrogen can be obtained by various types of electrical discharges. The temperature and pressure range for diamond synthesis are 900-1000°C and 5-10 Torr. The morphology of diamond single crystals and diamond film is very sensitive to CH_4 concentrations in H_2, pressure, temperature and physical location of substrate into the plasma. A diamond crystal grown by PACVD is shown in Figure 3.

Similar to diamond, cubic boron nitride has seen a surge in research interest in synthesizing C-BN film because of its unique combination of high hardness, physical, mechanical and optical properties. Salon, et.al, [44] have employed 30 KeV N_2^+ ion beam bombardment to synthesize Cubic Boron Nitride film. Shamfield, et.al [45] and many others [46,47] have reported producing CBN films by R.F. excited NH_3 plasma using borozine, by electron beam evaporation of boron in an NH_3 plasma and by reactive diode sputtering techniques. In a combined study of Bunshah and Chopra [48,49] they have reported a unique process for synthesizing CBN films which might be of significant importance, because of its potential applications in micro and optoelectronics as well as in solid lubrications and hard coatings. They claimed to have used boric acid, a non-toxic material, as opposed to toxic materials such as borozinc, diborane, etc. used by others for the preparation of CBN film. Bunshah, et. al., [50-52] have termed their process the Activated Dissociation Reduction Reaction Process (ADRRP) and it involves evaporation of boric acid in an NH_3 plasma. They have further claimed to deposit CBN at 400°C with the deposition rate up to 1500 Å/min. Finally, they could not detect any hexagonal phase in CBN film which shows characteristic absorption at 6.8 and 121.5 MM corresponding to B-N and N-B-N bending vibrations.

Concluding Remarks

Plasma has proven to be a very important tool in thin films deposition technologies. As a result, the old established processes for the surface modification such as hot dip molten, method diffusion, vacuum evaporation and electroplating have been replaced by the relatively new plasma assisted processes such as PACVD, Laser, Ion Plating, PAPVD and Ion Beam, to name a few. However, it is clear from the discussion in this paper, that there is a need to develop plasma assisted deposition processes where most of the process parameters can be independently controlled. Although plasma-substrate interactions result in desirable substrate chemistry changes, i.e., heating, surface cleaning and sputtering of loosely bonded species, there is a need to separate plasma from the substrate for obtaining film purity, lowering the deposition temperature and minimizing the bombardment induced film damages. Therefore, considerable research efforts and financial resources are needed to further understand plasma physics and the various species present in the plasma and their efforts on the reaction kinetics to obtain a desired deposited film at a much enhanced rate of deposition.

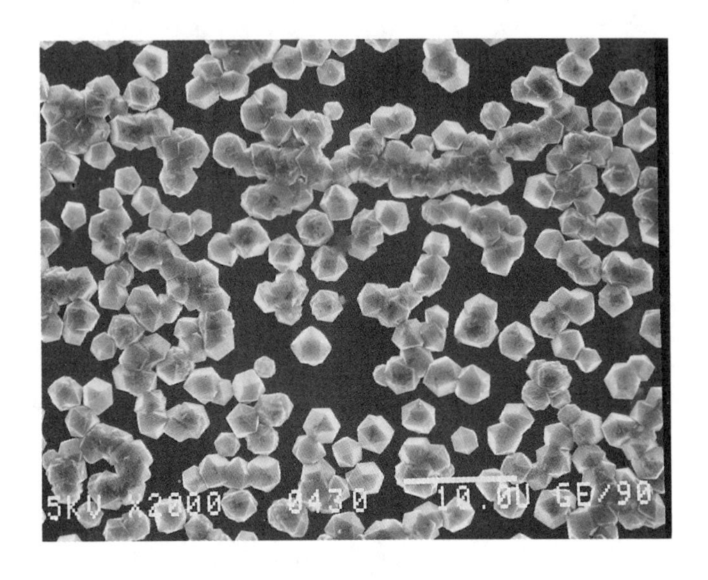

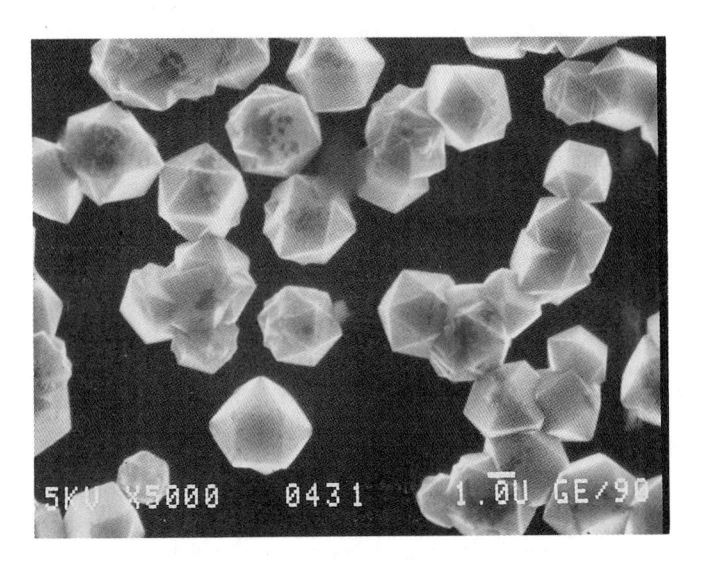

Figure 3 Micrographs of Diamonds Grown by Hot Filament CVD Technique

References

1. A.J. Thornton, in *Depositon Technologies for Films and Coatings*, Noyes, M.J. edited by R.F. Bunshah, p.232 (1982).

2. C. Deshpandey and R.F. Bunshah, Invited Paper presented in 1988 MRS Meeting, Boston, U.S.A.

3. R.F. Bunshah, Thin Solid Films (107) 21, (1983).

4. R.F. Bunshah in *Deposition Technologies for Films and Coatings*, Noyes, NJ, edited by R.R. Bunshah, p.5 (1982).

5. L. Holland in Vacuum Deposition of Thin Films, Chapman and Hall, London, p.62, (1956).

6. J.L. Vossen and J.J. Cuomo, in *Thin Film Processes* editd by J. Vossen and W. Kern (Academic, Nyew York, 1978).

7. C. Deshpandey, Doctoral thesis, University of Sussex, Falmer, Brighton, United Kingdom, 1981.

8. J.A. Thornton, in *Deposition Technologies for Films and Coatings*, edited by R.F. Bunshah (Noyes, New Jersey, 1982, p.232.

9. W. Westwood, Prog. Surf. Sci. **7**, 71 (1976).

10. J. Heller, Thin Solid Films, **17**, 163 (1973).

11. K.G. Geraghty and L.F. Donaghey, J. Electrochem. Soc. **123**, 1201 (1976).

12. L.F. Donaghey and K.G. Geraghty, Thin Solid Films **38**, 271 (1976).

13. S. Maniv and W.D. Westwood, J. Appl. Phys. **51**, 718 (1980).

14. S. Maniv and W.D. Westwood, J. Vac. Sci. Technol **3**, 743 (1980).

15. A.M. Stirlig and W.D. Westwood, J. Appl. Phys. **41**, 742 (1970).

16. A.J. Stirlig and W.D. Westwood, Thin Solid Films **7**, 1 (1971).

17. T. Abe and T. Yamashina, Thin Solid Films **30**, 19 (1974).

18. F. Shinoki and A. Itoh, J. Appl. Phys. **46**, 381 (1975).

19. B.R. Natarajan, A.H. Etoukhy, J.E. Green, And T.L. Barr, Thin Solid Films **69**, 201 (1980).

20. B. R. Natarajan, A.H. Ethoukhy, J.E. Green, and T.L. Barr, Thin Solid Films, **69**, 217 (1980).

21. C. Deshpandey and L. Holland, Thin Solin Films **97**, 265 (1982).

22. C. Deshpandey and L. Holland, *International Conference on Metal Coatings* (Iron and Steel Institute of Japan, Tokyo, 1982), p.276.

23. A.R. Nyiesh and L. Holland, J. Vac. Sci. Technol. **20**, 1389 (1982).

24. A.R. Nyiesh and L. Holland, Vacuum **31**, 371 (1981).

25. L. Holland, Thin Solid Film **86**, 227 (1981).

26. A.T. Bell in *Techniques and Applications of Plasma Chemistry*, edited by J.R. Hollahan and A.T. Bell, p.31, Wiley and Sons, NY (1974).

27. F.J. Kampas, in *Semiconductors and Semi-Metals*, edited by J.I. Pankov, Academic, NY, vol. 21, p. 159, (1984).

28. J.A. Thornton, Thin Solid Films, **107,** p.3, (1983).

29. L. Holland, Surface Technologies, **11**, p.145, (1980).

30. R.F. Bunshah and A.C. Raghuram, J. Vac. Sci. Tech. **9**, p.1385 (1972).

31. Ibrd, **9** p.1389 (1972).

32. R.C. DeVries in Am. Rev. Mat. Sci. **17**, p. 161 91987).

33. D.S. Whitmell and R.F. Williamson, Thin Solid Films, **35**, p.255 (1976).

34. L. Holland and S.M. Ojha, Thin Solid Films, **38**, L17 (1976).

35. D.A. Anderson, Phil. Mag., **35**, p.17 (1977).

36. S. Berg and L.P. Anderson, Thin Solid Films, **58**, p.117 (1979).

37. A.R. Badzian and R.C. DeVries, Mat. Res. Bull., Vol.23, p.385 (1988).

38. A.R. Badzian, T. Badzian, R. Roy, R. Messier and K.E. Spear, Mat. Res. Bull., **23**, p.531 (1988).

39. A.R. Badzian, B. Simonton, T. Badzian, R. Messier, K.E. Spear and R. Roy, SPIE, vol. 683, p.127 (1986).

40. S. Aisenberg and R. Chabot, J. Appl. Phys. **39**, P.2915 (1968).

41. E.G. Spencer, H.P. Schmidt, D.C. Joy and F.J. Sawsadone, Appl. Phys. Lett. **29**, p.118 (1976).

42. G. Gantherin and Chr. Weismental, Thin Solid Film, **50**, p.135, (1978).

43. S. Matsumoto, Y. Sato, M. Tsutsumi and N. Setaka, J. Mat. Sci., **17**, p.3106, (1982).

44. M. Salon, and F. Fujimoto, Japan J. Appl. Plug. Part 2, 22, 171, (1983).

45. S. Shamfield and R. Wolfson, J. Vac. Sci. Tech. (A) 1, 323, (1983).

46. C. Weissmantel, J. Vac. Sci. Tech. 18, 19, (1981).

47. P. Lin, C. Deshpandey, H.J. Doerr, R.F. Bunshah, K.L. Chopra and V.D. Vankar, Thin Solid Films, (1987).

48. K.L. Chopra, V. Agrawal, V.D. Vankar, C.V. Deschpandey, R.F. Bunshah, Thin Solid Films, 126, 307, (1985).

49. R.F. Bunshah, K.L. Chopra, C.V. Deshpandey and V.D. Vankar, P.T.N. 4714625 (1987) USA

50. K.L. Chopra, V. Agrawal, V.D. Vankar, C.V. Deshpandey, R.F. Bunshah, Thin Solid Films, vol.126, p.307, (1985).

51. P. Lin, C. Deshpandey, H.J. Doerr, R.F. Bunshah, K.L. Chopara, V.D. Vankar, Thin Solid Film, vol.153, p.487, (1987).

52. R.F. Bunshah, in *1*, Noyes Publications, p.14, (1982).

PLASMA SYNTHESIS OF THIN FILMS AND MULTILAYERS

WITH TAILORED ATOMIC MIXING

Ian Brown

Lawrence Berkeley Laboratory
University of California
Berkeley, CA 94720

Abstract

A method for the plasma synthesis of metallic and composite thin films with atomically mixed interfaces is described. Surface structures can be fabricated including films of metals and alloys, compounds including ceramics, and tailored multilayers. The added species can be energetically implanted below the surface or built up as a surface film with an atomically mixed interface with the substrate. A vacuum arc metal plasma is used, and by adding a gas to the plasma region compound films can be formed also. We have demonstrated the method by synthesizing a number of metallic, oxide and nitride films with different kinds of structures including metallic films atomically mixed to the substrate, multilayers with atomic stitching between chosen layers and to the substrate, alumina films atomically bonded to a steel substrate, and others. Here we outline the technique and the results that we've obtained.

Work supported by the Electric Power Research Institute under RP2426-27 and by the U.S. Department of Energy, Office of Basic Energy Sciences, Advanced Energy Projects, under Contract No. DE-AC03-76SF00098.

Plasma Synthesis and Processing of Materials
Edited by Kamleshwar Upadhya
The Minerals, Metals & Materials Society ©1993

Other features can be added to the basic technique described above, such as multiple metal plasma guns, perhaps of different metal species; variation of pulse length of the metal plasma pulse; phasing of the high voltage bias pulse (implantation phase) with respect to the plasma pulse (low energy plasma deposition phase); and all of these parameters can be tailored throughout the duration of the surface processing operation to fabricate a wide range of surface structures at the atomic level. By adding a gas flow to the deposition and implantation processes, the variety of films that can be formed is greatly expanded, and not only metallic films but also films of compounds including ceramic oxides, for example, can be formed. The technique can be used to form surface metal film structures that are of importance to a number of different areas, such as for the fabrication of metallic multilayers with novel mechanical properties that have super-strong adhesion to the substrate metal. Ceramic films can be formed on metal surfaces with atomically mixed transition zones and very good adhesion properties. The bonding transition zone between substrate and film can in principle be tailored to provide a good match between substrate and film properties, for example thermal expansion or lattice constant.

We have illustrated the principles of the method in a number of different experiments. As a demonstration of the basic concept, yttrium was deposited onto and recoil mixed into silicon with an atomically mixed transition zone extending into the substrate for about 500 Å. A thin film of platinum was formed on and mixed into an aluminum substrate. We formed a 6-layer multilayer structure of yttrium and titanium on a silicon substrate, in which the first and last layers were atomically mixed to the material below. We have also demonstrated the formation of oxide and nitride films with mixed transition zones of order 1000 Å thick, for example we've synthesized an alumina film on a steel substrate with atomic mixing of both the aluminum and the oxygen into the steel. In this latter case we've also carried out some scratch-adhesion testing and we've shown that the atomic mixing of the film to the substrate does indeed provide excellent adhesion characteristics to the film/substrate interface.

Experimental Approach

A convenient way of producing metal plasma is by means of a vacuum arc [13]. This kind of plasma discharge takes place between metallic electrodes in a high vacuum environment and is a prolific producer of dense metal plasma. Vacuum arc plasma sources have been used for the deposition of metallic thin films [14-16], and for the formation of metallic coatings and TiN by Physical Vapor Deposition (PVD) [17,18]. Industrial arc source deposition equipment is readily available on the market. The vacuum arc is commonly also called a 'cathodic arc', especially when used in a dc mode. Cathodic arc PVD facilities are large pieces of equipment and are designed to apply metallic coatings to large substrate areas; the titanium nitriding of cutting tools and other components in a large batch processing mode is a typical application. In the experiments described here, a pulsed vacuum arc plasma source was used to generate the required flux of metal plasma. The dense metal plasma plumes away from the cathode at which it is created, through an annular anode, and toward the substrate. The plasma guns are repetitively pulsed and a thin film can be deposited over an area of several square centimeters in a time of several minutes. The instantaneous deposition rate (ie, during the plasma pulse) is typically of order several hundred up to 1000 Å/sec, and the time-averaged rate is of order 0.1 to 1 Å/sec. The properties of the plasma generated by this kind of plasma gun have been extensively studied as part of the LBL vacuum arc ion source program [19-21]. The source works with virtually all solid metals of the periodic table; we have operated with 50 different metallic elements [22,23] - Li, C, Mg, Al, Si, Ca, Sc, Ti, V, Cr, Mn, Fe, Co, Ni, Cu, Zn, Ge,Sr, Y, Zr, Nb, Mo, Pd, Ag, Cd, In, Sn, Sb, Ba, La, Ce, Pr, Nd, Sm, Gd, Dy, Ho, Er, Tm, Yb, Hf, Ta, W, Ir, Pt, Au, Pb, Bi, Th, U - and a range of alloys and compounds [24]. The source is efficient and it can be scaled up. It operates in a high vacuum environment; no support gas is required. We've made a number of different embodiments of vacuum arc plasma gun, from tiny, sub-miniature versions up to large, water-cooled dc versions. A multiple-cathode version has also been made, whereby one can switch from one cathode species to another very simply and without breaking vacuum. A photograph of one version of plasma gun that we've used is shown in Figure 1. The experiments were carried out in a vacuum vessel pumped cryogenically to a pressure of about 1 x 10^{-6} Torr.

Other features can be added to the basic technique described above, such as multiple metal plasma guns, perhaps of different metal species; variation of pulse length of the metal plasma pulse; phasing of the high voltage bias pulse (implantation phase) with respect to the plasma pulse (low energy plasma deposition phase); and all of these parameters can be tailored throughout the duration of the surface processing operation to fabricate a wide range of surface structures at the atomic level. By adding a gas flow to the deposition and implantation processes, the variety of films that can be formed is greatly expanded, and not only metallic films but also films of compounds including ceramic oxides, for example, can be formed. The technique can be used to form surface metal film structures that are of importance to a number of different areas, such as for the fabrication of metallic multilayers with novel mechanical properties that have super-strong adhesion to the substrate metal. Ceramic films can be formed on metal surfaces with atomically mixed transition zones and very good adhesion properties. The bonding transition zone between substrate and film can in principle be tailored to provide a good match between substrate and film properties, for example thermal expansion or lattice constant.

We have illustrated the principles of the method in a number of different experiments. As a demonstration of the basic concept, yttrium was deposited onto and recoil mixed into silicon with an atomically mixed transition zone extending into the substrate for about 500 Å. A thin film of platinum was formed on and mixed into an aluminum substrate. We formed a 6-layer multilayer structure of yttrium and titanium on a silicon substrate, in which the first and last layers were atomically mixed to the material below. We have also demonstrated the formation of oxide and nitride films with mixed transition zones of order 1000 Å thick, for example we've synthesized an alumina film on a steel substrate with atomic mixing of both the aluminum and the oxygen into the steel. In this latter case we've also carried out some scratch-adhesion testing and we've shown that the atomic mixing of the film to the substrate does indeed provide excellent adhesion characteristics to the film/substrate interface.

Experimental Approach

A convenient way of producing metal plasma is by means of a vacuum arc [13]. This kind of plasma discharge takes place between metallic electrodes in a high vacuum environment and is a prolific producer of dense metal plasma. Vacuum arc plasma sources have been used for the deposition of metallic thin films [14-16], and for the formation of metallic coatings and TiN by Physical Vapor Deposition (PVD) [17,18]. Industrial arc source deposition equipment is readily available on the market. The vacuum arc is commonly also called a 'cathodic arc', especially when used in a dc mode. Cathodic arc PVD facilities are large pieces of equipment and are designed to apply metallic coatings to large substrate areas; the titanium nitriding of cutting tools and other components in a large batch processing mode is a typical application. In the experiments described here, a pulsed vacuum arc plasma source was used to generate the required flux of metal plasma. The dense metal plasma plumes away from the cathode at which it is created, through an annular anode, and toward the substrate. The plasma guns are repetitively pulsed and a thin film can be deposited over an area of several square centimeters in a time of several minutes. The instantaneous deposition rate (ie, during the plasma pulse) is typically of order several hundred up to 1000 Å/sec, and the time-averaged rate is of order 0.1 to 1 Å/sec. The properties of the plasma generated by this kind of plasma gun have been extensively studied as part of the LBL vacuum arc ion source program [19-21]. The source works with virtually all solid metals of the periodic table; we have operated with 50 different metallic elements [22,23] - Li, C, Mg, Al, Si, Ca, Sc, Ti, V, Cr, Mn, Fe, Co, Ni, Cu, Zn, Ge,Sr, Y, Zr, Nb, Mo, Pd, Ag, Cd, In, Sn, Sb, Ba, La, Ce, Pr, Nd, Sm, Gd, Dy, Ho, Er, Tm, Yb, Hf, Ta, W, Ir, Pt, Au, Pb, Bi, Th, U - and a range of alloys and compounds [24]. The source is efficient and it can be scaled up. It operates in a high vacuum environment; no support gas is required. We've made a number of different embodiments of vacuum arc plasma gun, from tiny, sub-miniature versions up to large, water-cooled dc versions. A multiple-cathode version has also been made, whereby one can switch from one cathode species to another very simply and without breaking vacuum. A photograph of one version of plasma gun that we've used is shown in Figure 1. The experiments were carried out in a vacuum vessel pumped cryogenically to a pressure of about 1 x 10^{-6} Torr.

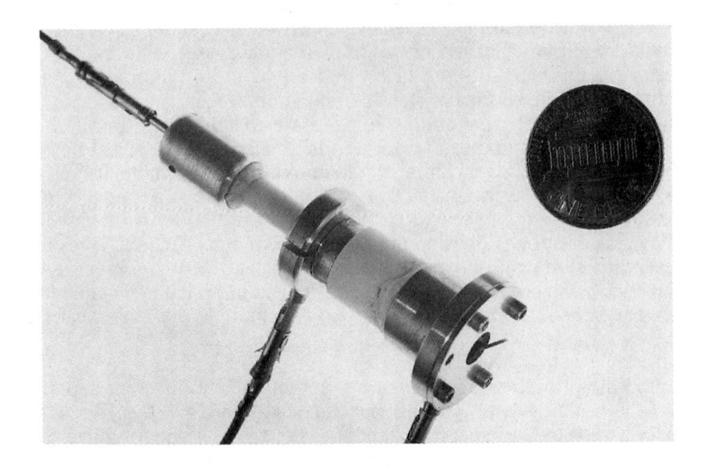

Figure 1 - One version of pulsed vacuum arc plasma gun.

Along with the metal plasma that is generated by the vacuum arc there is also produced a flux of macroscopic droplets of size typically in the broad range 0.1 - 10 microns [25-27]. We have examined the films under SEM and a 'macroparticle' contamination is generally present, the degree of this contamination depending on the particular cathode material used, especially its melting point. The origin of the macroparticles is at the cathode spots, where they are produced by the intense heating of the cathode material beneath the spot. The volume of molten material formed is less for higher melting point materials, and the macroparticle contamination is observed to be less for cathode materials of higher melting point. For some applications, useful films can be made by proper selection of cathode material. For those applications for which the macroparticle generation is severe or for which truly 'macro-free' surfaces are required, a magnetic filter can be used, as has been investigated and described by several workers. One simple configuration is a curved 'magnetic duct' which stops line-of-sight transmission of macroparticles while allowing the transmission of plasma by virtue of an axial magnetic field which ducts the plasma through the filter [28-33]. A photograph showing a vacuum-arc-produced platinum plasma exiting from one of our magnetic ducts and depositing onto a silicon substrate is shown in Figure 2.

Figure 2 - Macroparticle-free platinum plasma streaming from a magnetic duct and depositing onto a silicon substrate.

In the work that we've done to demonstrate our metal plasma immersion processing technique, we've used one or more metal plasma guns set up in a configuration as indicated in Figure 3. Previous work [22,23] has shown that the vacuum arc metal plasm is multiply ionized with a mean charge state that lies between 1 and 3 depending on the metal species used; thus, when implanted, the mean ion energy is greater than the applied bias pulse voltage by this factor. Prior to implantation (repetitive application of the high voltage pulse), the plasma ion current collected by the substrate during each pulse was measured by biasing the substrate to about -200 volts and measuring the ion saturation current in the usual way [34]. The ion saturation current pulse, ie, the plasma ion current density, was typically several microseconds wide and several amperes in magnitude. For synthesizing compound and ceramic films, gas was injected into the vicinity of the metal plasma and the depositing film. The gas then participates in the implantation and/or deposition phases of the processing.

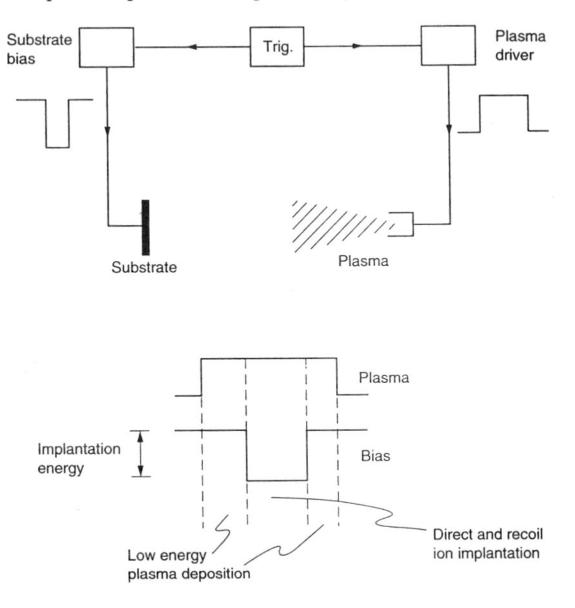

Figure 3 - Simplified schematic of the experimental configuration used.

Experimental Results

Several different experiments showing the different kinds of surfaces that can be formed by our technique have been carried out. Each involves the synthesis of a film or multilayer structure that is atomically mixed to the substrate. A review of these experiments and their results is summarized briefly in the following.

Metal Plasma Immersion Ion Implantation - Yttrium Implanted into Silicon

The plasma gun was operated with yttrium and the silicon substrate was repetitively pulse biased with -30 kV, 1μs pulses timed for the maximum of the plasma pulse. The equivalent time-averaged ion implantation current was several tens of microamperes onto the substrate of area several square centimeters. The mean ion charge state of the vacuum-arc-produced yttrium plasma was 2.3 [22,23], and thus the 30 kV voltage pulse corresponded to a mean ion energy of 70 keV. Another sample was prepared under identical conditions except that the high voltage implantation pulse was not applied, thus providing a non-implanted sample for comparison. The samples were analyzed by 2.0 MeV He^+ Rutherford Backscattering Spectrometry (RBS), and the results are shown in Figure 4. The RBS resolution is indicated by the Gaussian-shaped profile obtained when pulse biasing is not applied. With pulse biasing the depth profile extends below the surface and has a shape that is qualitatively as expected for a combination of conventional ion implantation, recoil implantation, and surface deposition. A nominal depth of the implanted region (half width of the profile with an ad-hoc correction for the RBS resolution) is approximately 500 Å, which is comparable to the TRIM-calculated [35] range for 70 keV Y into Si of 470 Å. The RBS-measured dose was 1.0 x 10^{16} atoms/cm^2, in agreement with the dose expected from the accumulated number of pulses. This simple application of the method thus confirms that the technique works fundamentally as expected.

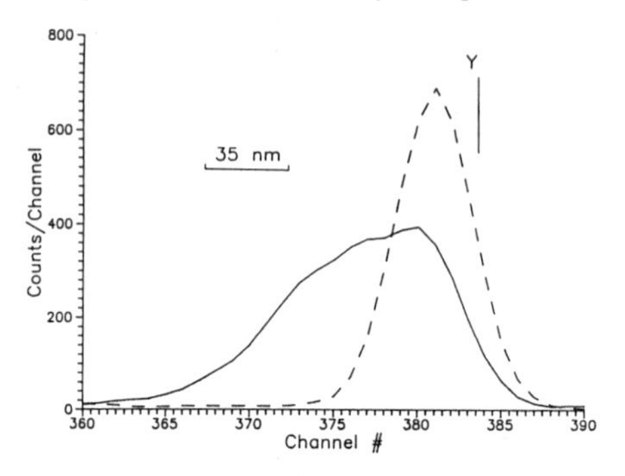

Figure 4 - Rutherford Backscattering spectra for Y into Si. The dashed curve is from the sample without pulse biasing and the solid curve from that with pulse biasing. The depth of implantation is approximately 500 Å and the dose is 1 x 10^{16} cm^{-2}.

Thin Film Atomically Mixed to Substrate - Platinum on Aluminum

The plasma gun was operated with a Pt cathode and the substrate was Al. The pulse biasing (piii) phase was followed by a long-pulse deposition phase so as to build up a film of thickness about 0.25 μm. Thus we produced a simple platinum film that is atomically mixed to the aluminum substrate. RBS data of this structure are shown in Figure 5.

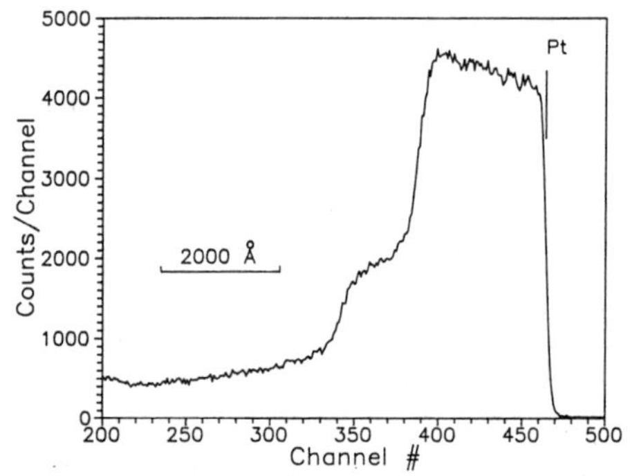

Figure 5 - Rutherford Backscattering spectrum of Pt film atomically mixed into Al substrate.

Multilayer Structure with Atomic Mixing - Yttrium/Titanium on Silicon

Two plasma guns were used, one with an yttrium cathode and the other with titanium. The sequence of operations was: yttrium was implanted into the substrate at an energy of 70 keV and a dose of 1 x 10^{16} atoms/cm^2, followed by a low energy (about 50 eV; no high voltage pulse applied) deposition of yttrium to build up a layer of thickness several hundred Angstroms (the pulse length of the yttrium plasma pulse was increased from 2 μs up to 250 μs for this phase); then successive layers of Ti-Y-Ti-Y-Ti were added, each of several hundred Angstroms thickness; the final Ti layer was started out by implanting Ti into the underlying Y layer at an energy of about 50 keV and a dose of about 1×10^{15} atoms/cm^2, then followed by the low energy, longer pulse length part of the final Ti phase. The RBS data are shown in Figure 6. The multilayer structure is evident; the layers have a thickness of approximately 400 Å, corresponding to a deposited particle density of approximately 1.4×10^{17} atoms/cm^2 for Y and 2.5×10^{17} atoms/cm^2 for Ti.

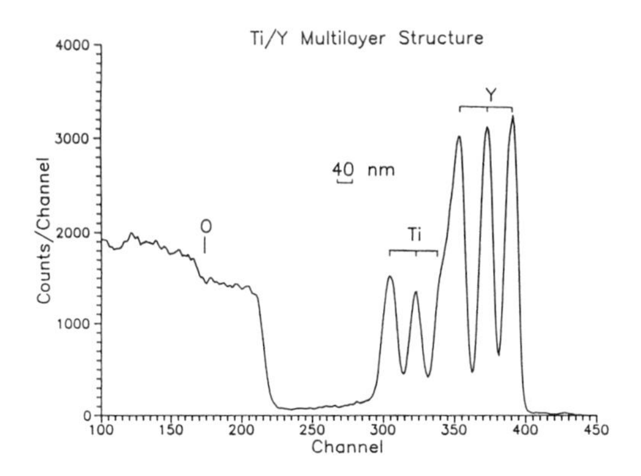

Figure 6 - RBS spectrum of a Ti-Y multilayer structure on Si with atomic mixing at the first Si-Y and the final Ti-Y interfaces.

Ceramic Film Atomically Mixed to Metallic Substrate - Alumina on Steel

The plasma gun was operated with an aluminum cathode and the substrate was polished stainless steel. An appropriately designed gas jet arrangement was used to inject oxygen into the plasma/substrate interaction region. The metal plasma immersion processing technique was applied so as to form an atomically mixed region, followed by a film of thickness about 0.2 μm. With some judicious selection of the various operational parameters, we were able to synthesize a stoichiometric film of alumina with mixing into the stainless steel substrate of both the aluminum and oxygen down to a depth of order 1000 Å. Auger data showing this depth profile are shown in Figure 7. Note that the close-to-stoichiometric Al:O ratio is maintained throughout

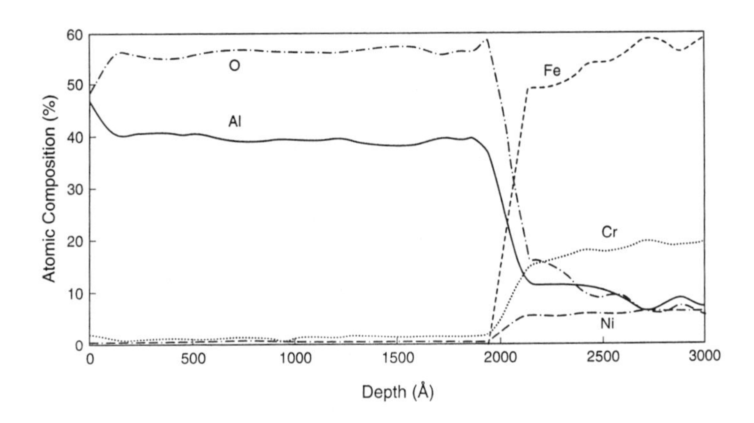

Figure 7 - Auger depth profile of Al_2O_3 film on stainless steel with atomic mixing at the interface, showing ~1000Å thick intermixed zone.

most of the film. To better reveal the atomic mixing of the Al and O into the steel substrate, the Auger data have been cast into a different form in Figure 8. Here we view the "3-dimensional" Auger profiles from the inside of the material looking toward the interface; the slow blending of both the Al and O profiles into the substrate (for ~1000Å as indicated by Figure 7) is manifest.

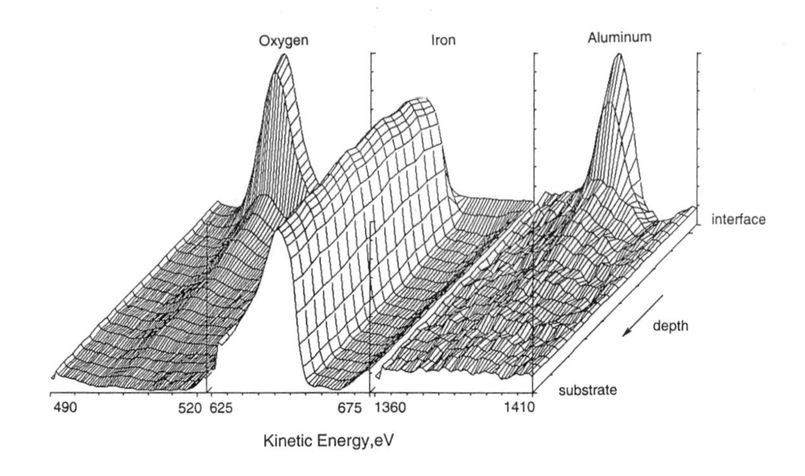

Figure 8 - "3D" Auger profiles showing the atomic mixing of the Al and O in the steel substrate.

Using an inexpensive, 'home-made' scratch/adhesion testing device in which the sample is moved uniformly below a precisely weighted diamond stylus using a milling machine [36], we have compared the scratch/adhesion performance of alumina films on steel, both made identically (at the same time) except for the metal plasma immersion processing and consequent atomic mixing into the substrate for one of them. SEM photographs of the scratch marks made are shown in Figure 9. One can see that the film which has been simply deposited has lifted away from the substrate whereas the atomically mixed film has maintained its integrity with the substrate.

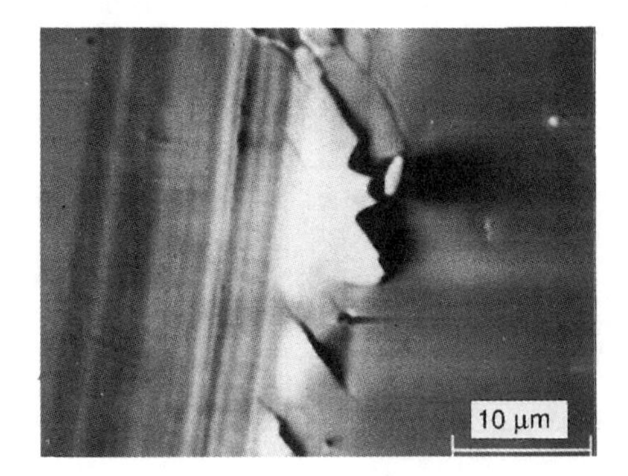

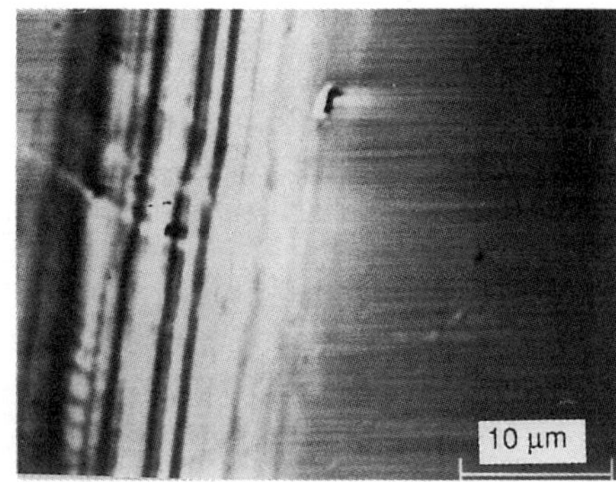

Figure 9 - SEM photographs showing the scratch tracks made in a 2000 Å thick Al_2O_3 film on a steel substrate without interface mixing (left), and with the film-substrate interface atomically mixed (right).

Conclusion

The metal plasma immersion surface processing technique provides a means for doing surface modification and materials synthesis with a range of features and properties that could be valuable in many ways. The method is fundamentally simple and the hardware required is not exotic or difficult to make. We've carried out a number of experiments that demonstrate that the technique works and that the surfaces created possess novel and interesting properties. It is clear from these experiments and results that the surface structures synthesized can be tailored over a very wide range of parameter space; the demonstration examples chosen were arbitrary. Note that the tests reported here were our first tests; the results could be improved upon in many ways by further experimental iterations of the process so as to home-in on the optimum parameters for the synthesized structures. For example, the recoil-implanted zone could be better matched to the film above it by more precisely tailoring the implantation dose accumulated during the pulse biasing phase of the process, and the range or depth of the mixed zone can be tailored simply by the bias voltage magnitude. There is much more work that could be done to complete the picture outlined here.

One can think of many additional features that could be added to create better, or bigger, or quicker variations on the metal plasma immersion theme. For example the method can readily be scaled up to produce atomically bonded macroscopically thick coatings; the method is not at all limited to thin films. Similarly the scale-up could be to large substrate area and high throughput. One can envision a computer controlled processing system, perhaps with feedback from film thickness monitors, whereby it would be possible to build up highly complex structures of metals, ceramics, and other composites and compounds, having films and multilayers with sharp or graded interfaces between layers and to the substrate. The bonding transition zone between substrate and film and between different layers of a multilayered structure can be tailored widely to provide an optimum match of film properties, for example thermal expansion or lattice constant.

The results demonstrate that this metal plasma immersion surface processing technique can be used to form surface layers that could be of relevance to a number of fields. The hardware required is neither complicated nor expensive, and the method can be scaled up much beyond the preliminary experiments described here.

Acknowledgements

The work described here has been accomplished as a group effort. I am greatly indebted to my scientific colleagues including Drs. Andre and Simone Anders, Igor Ivanov, Xiang Yao and Kin-Man Yu. The mechanical design and fabrication of the sources and other equipment was done by a team led by Bob MacGill and including Mike Dickinson and Bob Wright, and the electrical and electronics support was provided by Jim Galvin, Jan deVries, Bud Leonard and Mark Rickard. This work was supported by the Electric Power Research Institute under Award number RP2426-27 and by the Department of Energy, Office of Basic Energy Sciences, Advanced Energy Projects, under Contract No. DE-AC03-76SF00098.

References

1. See, for instance, "Ion Implantation and Plasma Assisted Processes", edited by R. F. Hochman, H. Solnick-Legg and K. O. Legg, (ASM, Ohio, 1988).
2. "Plasma Processing and Synthesis of Materials", edited by D. Apelian and J. Szekely, Mat. Res. Soc. Symp. Proc. Vol 98, (MRS, Pittsburgh, 1987).
3. G. Dearnaley, Nucl. Instr. and Meth. B50, 358 (1990).
4. M. Iwaki, Critical Rev. in Solid State and Mat. Sci., 15, 473 (1989).
5. L. E. Rehn and P. R. Okamoto, Nucl. Instr. and Meth. B39, 104 (1989).
6. G. K. Wolf and W. Ensinger, Nucl. Instr. and Meth. B59/60, 173 (1991).
7. J. R. Conrad, J. L. Radtke, R. A. Dodd, F. J. Worzala and N. C. Tran, J. Appl. Phys. 62, 4591 (1987).

8. J. T. Scheuer, M. Shamim and J. R. Conrad, J. Appl. Phys. 67, 1241 (1990).
9. J. Tendys, I. J. Donnelly, M. J. Kenny and J. T. A. Pollock, Appl. Phys. Lett. 53, 2143 (1988).
10. H. Wong, X. Y. Qian, D. Carl, N. W. Cheung, M. A. Lieberman, I. G. Brown and K. M. Yu, Mat. Res. Soc. Symp. Proc. 147, 91, (MRS, Pittsburgh, 1989).
11. X. Y. Qian, H. Wong, D. Carl, N. W. Cheung, M. A. Lieberman, I. G. Brown and K. M. Yu, 176th Electrochemical Society Meeting, Hollywood, Fla, October 15-20, 1989.
12. X. Y. Qian, M. H. Kiang, J. Huang, D. Carl, N. W. Cheung, M. A. Lieberman, I. G. Brown, K. M. Yu and M. I. Current, Nucl. Instrum. Meth. Phys. Res. B55, 888
13. "Vacuum Arcs - Theory and Application", edited by J. M. Lafferty, Wiley, New York, 1980.
14. R. L. Boxman, S. Goldsmith, S. Shalev, H. Yaloz and N. Brosh, Thin Solid Films 139, 41 (1985).
15. D. M. Sanders, "Review of Ion Based Coating Processes Derived from the Cathodic Arc", J. Vac. Sci. Tech. A7, 2339 (1989).
16. X. Godechot, M. B. Salmeron, D. F. Ogletree, J. E. Galvin, R. A. MacGill, K. M. Yu and I. G. Brown, "Thin Film Synthesis using Miniature Pulsed Metal Vapor Vacuum Arc Plasma Guns", Mat. Res. Soc. Symp. Proc. 190, 95 (1991).
17. C. Bergman, in "Ion Plating and Implantation", edited by R. F. Hochman, American Society for Metals, USA, 1986. (Proceedings of the ASM Conference on Applications of Ion Plating and Implantation to Materials, June 3-5, 1985, Atlanta, GA).
18. P. A. Lindfors, loc. cit. [17].
19. I. G. Brown, J. E. Galvin, and R. A. MacGill, Appl. Phys. Lett. 47, 358 (1985).
20. I. G. Brown, in "The Physics and Technology of Ion Sources", I. G. Brown editor, (Wiley, N.Y., 1989).
21. I. G. Brown, Rev. Sci. Instrum. 63, 2351 (1992).
22. I. G. Brown, B. Feinberg, and J. E. Galvin, J. Appl. Phys. 63, 4889 (1988).
23. I. G. Brown and X. Godechot, IEEE Trans. Plasma Sci. PS-19, 713 (1991).
24. J. Sasaki and I. G. Brown, J. Appl. Phys. 66, 5198 (1989).
25. D. T. Tuma, C. L. Chen and D. K. Davies, J. Appl. Phys. 49, 3821 (1978).
26. J. E. Daalder, Physica 104C, 91 (1981).
27. I. I. Aksenov, I. I. Konovalov, E. E. Kudryavtseva, V. V. Kunchenko, V. G. Padalka and V. M. Khoroshikh, Sov. Phys. Tech. Phys. 29(8), 893 (1984).
28. I. I. Aksenov, V. A. Belous, V.G. Padalka and V. M. Khoroshikh, Sov. J. Plasma Phys. 4(4), 425 (1978).
29. V. A. Osipov, V. G. Padalka, L. P. Sablev and R. I. Stupak, Instrum. and Exp. Techniques 21(6), 173 (1978).
30. I. I. Aksenov, S. I. Vakula, V. G. Padalka, V. E. Strelnitski and V. M. Khoroshikh, Sov. Phys. Tech. Phys. 25(9), 1164, (1980).
31. V. S. Voitsenya, A. G. Gorbanyuk, I. N. Onishchenko and B. G. Safranov, Sov. Phys. - Tech. Phys. 9(2), 221 (1964).
32. I. I. Aksenov, A. N. Belokhvostikov, V. G. Padalka, N. S. Repalov and V. M. Khoroshikh, Plasma Physics and Controlled Fusion 28, 761 (1986).
33. J. Storer, J. E. Galvin and I. G. Brown, J. Appl. Phys. 66, 5245 (1989).
34. See, for instance, F. F. Chen in "Plasma Diagnostic Techniques", edited by R. H. Huddlestone and S. L. Leonard (Academic Press, N. Y., 1965).
35. J. F. Ziegler, J. P. Biersack and U. Littmark, in "The Stopping and Range of Ions in Solids", Vol 1, edited by J. F. Ziegler (Pergamon, N.Y., 1985).
36. R. A. MacGill, X. Y. Yao, R. A. Castro, M. R. Dickinson and I. G. Brown, “Poor Man’s Scratch Tester”, submitted to J. Vac. Sci. Tech.

SPUTTERING FOR SEMICONDUCTOR APPLICATIONS

Stephen M. Rossnagel
IBM T.J. Watson Research Center
PO 218, Yorktown Heights, NY 10598

Abstract

Sputtering, and in particular magnetron sputtering, has found broad acceptance and usage in the processing of semiconductors. Sputter deposition is routinely used for the deposition on metal layers for conductors as well as the reactive deposition of various oxides and nitrides for dielectric applications, diffusion and/or adhesion layers. Sputter etching is routinely used for etching various structures on surfaces, although more commonly it is combined with a chemical process in the form of reactive ion etching (RIE). More recently, variations on basic sputtering technology have been developed to address specific needs in the semiconductor industry as well as the coatings industry. These variations include alterations in magnetron design, such as the unbalanced magnetron or the rotating-magnet magnetron, as well as modifications to the transport process, such as collimation or post ionization. This paper will address some of the basic aspects of magnetron sputtering and then concentrate on some of these more recent developments.

INTRODUCTION

Magnetron sputtering became commonly used in the late 1970's with the advent of commercial sources and systems. A planar magnetron is the simplest design and is shown in Fig. 1. The primary consideration for the magnetron effect is that the ExB drift path for secondary electron emitted from the cathode forms a closed path. In the circular planar device, this path is circular with a radius of typically 1/2 the radius of the target. Other permutations on this geometry are the rectangular or "racetrack" magnetron (Fig. 2), the post magnetron, the hollow-post magnetron, nested-loop magnetrons, and so on.

Plasma Synthesis and Processing of Materials
Edited by Kamleshwar Upadhya

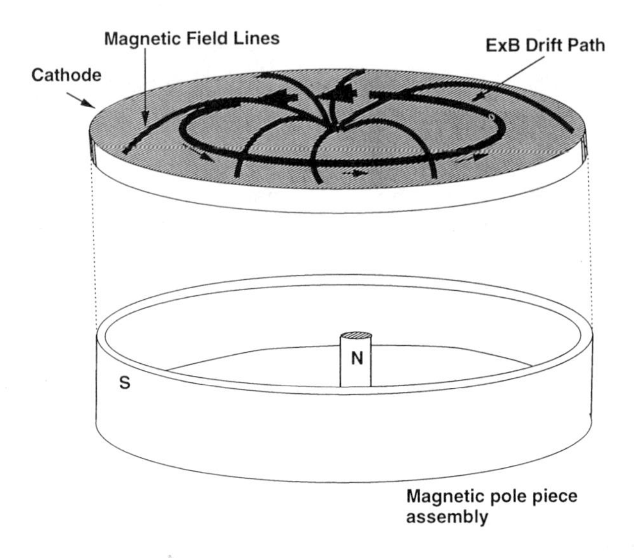

Figure 1. The circular planar magnetron

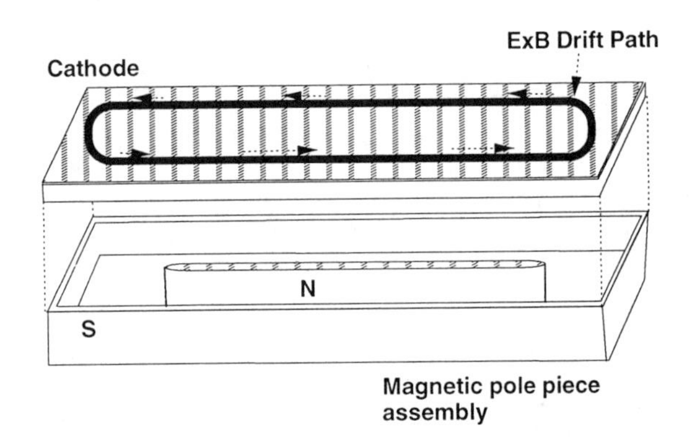

Figure 2. The rectangular, planar magnetron (racetrack magnetron).

The magnetron is a diode plasma device in which the cathode is typically powered negative of ground by several hundred volts, either DC or at 13.56 MHz. While rf-powered magnetrons are commonly used for some reactive sputtering applications, they will only be discussed briefly here. In general, rf magnetrons are less efficient, up to a factor of 2x, than their DC counterparts. The DC magnetron can be operated either with a specific anode electrode or with the chamber walls functioning as an effective anode. The plasma potential in this case it generally just slightly above the anode potential, and ions which cross the cathode sheath impact the cathode with an energy essentially equal to the discharge voltage.

The magnetron plasma operates in a space charge limited mode (1), but its discharge impedance is an unusual function of the current and voltage. The current-voltage relation can be described empirically by

$$I = kV^n$$

where I and V are the discharge current and voltage, and k and n are system specific, cathode and gas specific constants. Early work with magnetrons often resulted in tabulations of various values for k and n, which depended on the cathode material, the gas and gas pressure, and the origin of the cathode design. The unusual issue in this empirical relation is the exponent, n, which can be as high as 15 for common magnetrons, and has been observed to be as high as 75 for some systems. This indicates that the current will increase very rapidly for very small increases in discharge voltage. This is unlike all other diode plasmas, which have a more linear or at most quadratic dependance. This unusual behavior can be traced to strong variations in the sheath thickness with discharge current. In effect, the space charge limit is still in place, but the sheath thickness changes significantly due to the rapid increase in discharge ion density with increased current (1).

GAS DENSITY EFFECTS

In the early 1980's, an effect was observed by Dave Hoffman at Ford which was termed the "sputtering wind." This effect was a momentum-transfer based effect caused by the sputtering process which resulted in the formation of convection-like patterns in the gas flow in the sputtering chamber (2). Later work examined this effect in more detail. It was found that the gas density in the region near the sputtering cathode was strongly perturbed by the sputtering process (3). The gas density was strongly depleted as a function of increasing discharge current. The experiments used to measure this effect were quite simple, and consisted of a short sampling tube placed into the discharge region. The other end of the tube was connected to a capacitance manometer, and the real particle density could be calculated by using thermal transpiration theory.

Upon increasing the discharge current, a strong reduction in the gas density could be observed immediately in front of the cathode, although an identical manometer placed at the chamber wall (the normal situation in most experiments) did not observe any changes (Fig. 3). The effect was dramatic: the most severe reductions were of up to 85% of the initial density by using a discharge power of 2-3 kW in a 15 cm diameter magnetron. The magnitude of the effect varied. It was greatest for high sputter yield cathodes as well as for heavy sputtering gases.

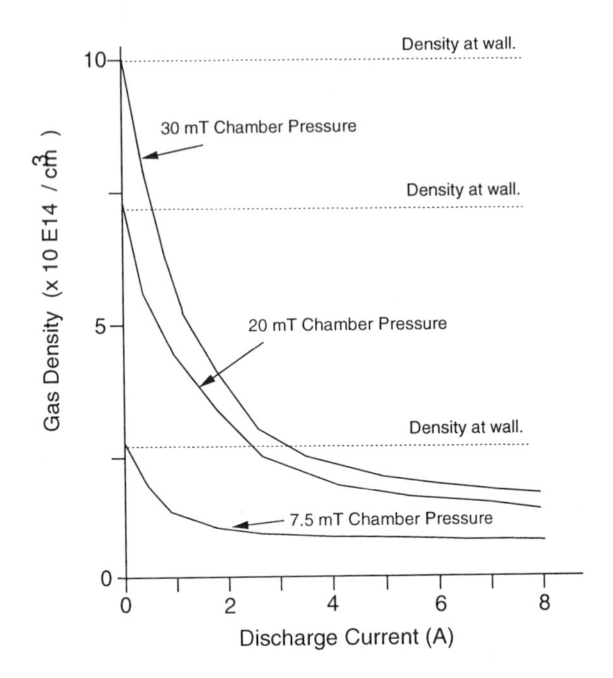

Figure 3. The gas density in the center region of a magnetron plasma, 5 cm from the cathode, compared to the density at the wall as a function of discharge current (15 cm dia cathode, Ar gas, Cu cathode).

The effect which causes this density reduction is the heating and rarefaction of the background gas by the kinetic energy transfer from the sputtered particles. To put this in perspective, most materials used for sputtering have a sputter yield of 1-2. If a discharge current of 5 Amperes is used to the cathode, then as many as 10 Amperes-equivalent of sputtered particles are emitted from the cathode surface. In terms of particle flux, this is over 140 SCCM of gas atoms. The sputtered atoms are emitted from the cathode with an average kinetic energy of from 5 to 35 eV, depending on the cathode choice. For common metal cathode choices, the net power put back into the gas phase due to the kinetic energy of the sputtered atoms is about 5% of the discharge power. This does not seem like much. However, the gas in front of the cathode is at rather low pressure and has very little heat capacity. In effect, the temperature of the local gas increases and its density decreases at the same time. This effect has been derived analytically and the density can easily be calculated as a function of the discharge current, the sputter yield, the thermal conductivity of the gas and the wall temperature. In effect, it is a 3-dimensional heat conduction process from the cathode surface to the wall regions.

The effect of the gas rarefaction is to alter a number of aspects of the system. The reduction in the number of gas atoms makes it more difficult to ionize the remaining background gas. The electron temperature increases slightly and the discharge voltage also increases to help heat the electrons. This gas rarefaction effect can be used to explain the differences in the discharge impedance curves (I-V Plots) as a function of gas species, pressure and cathode species. In effect, materials with high sputter yields have higher levels of gas rarefaction, and thus must increase their voltage faster with increasing current than low sputter yield cathodes. The same argument can be made in reverse for the thermal conductivity of the gas. Light gases, typically which have good thermal conductivity, show the gas rarefaction effect less because the gas atoms can move rapidly to conduct the heat away. A a result, the impedance of these discharges is less, and increases less rapidly with increasing current than discharges operated with heavy gas species which have lower thermal conductivity An extreme example of both of these effects is the sputtering of tungsten with helium. In this case, the yield is effectively zero and the thermal conductivity of the gas is the highest possible. No gas rarefaction is observed, and the discharge current increases very rapidly with discharge voltage, showing an effective exponent, n, of over 75.

SPUTTERED ATOM TRANSPORT

The goal of sputter deposition systems is the rapid deposition of high quality films on surfaces. The transport probability of the sputtered atoms is the likelihood that the sputtered atoms will deposit on the substrate plane, as opposed to other surfaces within the system. Once the atoms are sputtered from the cathode surface, they have an initial velocity and direction moving away from the cathode. This means, however, that some fraction of the sputtered atoms will be moving such that they strike the chamber walls or shields and as a result are not available to be deposited at the sample. In addition, if there are any significant number of sputtered atom-gas atom collisions (more than 2-3), the initial direction of the sputtered atom will be lost, and it will become part of the gas phase. In this case, it is no more likely to land on the substrate surface than it is to land on any other surface in the chamber, such as the cathode, the windows, etc.

The transport probabilities are sensitive to the geometry of the system, the presence of various shutters and fixtures in the system, the chamber pressure and the gas choice. As might be expected, increasing the distance between the cathode and the sample is likely to lead to a reduced deposition rate, as sputtered atoms can land on other surfaces in the chamber. It was found, however, that due to the rather awkward geometry of the magnetron system, little practical modeling, either analytical or monte-carlo, had been done to try to understand the transport process.

Experiments were undertaken to try to gauge the transport probability (4). In these studies, large numbers of samples were positioned around the chamber and the thickness of films deposited on them compared after identical runs. The number of film atoms on any given surface can then be compared with the original starting number of sputtered atoms to give the probability of transport. An example of the type of data gathered is shown in Fig. 4. In this case, the transport of sputtered Al is shown in Ne, Ar and Kr gases. The upper curves show the transport to the sample plane in front of the cathode, 5 cm away. The lower curves show the probability of redeposition back onto the cathode. These sets of curves begin to converge at higher pressure, and actually cross at larger throw distances. This means that in these cases, it is more likely for the sputtered atom to return to the cathode as it is for the atom to land on the sample, as desired.

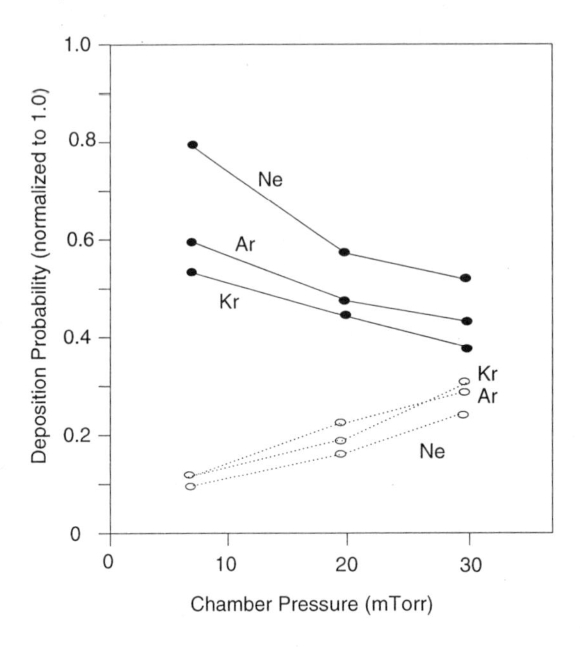

Figure 4. The transport probability for sputtered Al atoms as a function of chamber pressure and gas species for a 5 cm throw distance. The top 3 curves are for transport to the sample plane; the bottom three curves are for redeposition back onto the cathode.

COLLIMATED SPUTTERING

The semiconductor industry relies on the ability to pattern various circuit elements on Si wafer surfaces. Historically, this patterning was done with a technology known as "lift-off" in which a photoresist mask was developed on the wafer surface and metal deposited in a line-of-sight mode onto both the mask and the exposed areas in-between mask structures. Once the photoresist mask was taken away (lifted-off), the unwanted film was also removed, leaving the desired pattern on the surface.

Lift-off deposition required evaporation, because of the requirement of line-of-sight. An alternative developed in the 1980's was based on the selective removal of blanket metal films. The metal was deposited typically with magnetron sputtering, and then a photoresist mask developed on top of the blanket film. A plasma-based etching process (RIE or else physical sputtering) eroded the metal exposed through the mask. Once the mask was removed, the desired circuit elements, which were protected by the mask, remained. This

RIE-based etch-back process remains the industry standard for most metallization processes.

The common problem with either of these patterning technologies was that the resulting structures tended to be wide and not too thick. As circuit densities grow, though, there is pressure to reduce the scale of all active elements on the chip. This could not be done with line structures, because the patterning techniques did not allow the deposition of thick, narrow lines.

An alternative, recently developed, is the use of metal filled trench structures. As shown in Fig. 5, a deep, high aspect ration trench can be etched onto an insulating layer on the chip (quartz, typically). The next task is to fill up the trench entirely with metal, which usually results in additional metal deposition on the top surface of the wafer between the trenches. This metal is removed by a simple, physical polishing process, which grinds the surface back to the original level, leaving the metal line flush with the surface but buried. The result, now, is a high aspect ratio, fairly thick metal conductor, which uses fairly little surface area on the chip and has an intrinsically planar surface. This patterning technique is in common usage on programs such as high density memory DRAMS.

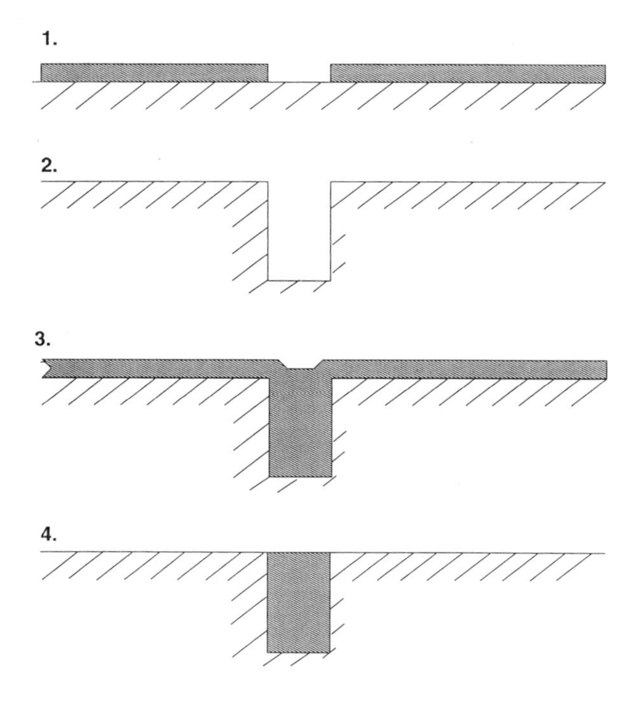

Figure 5. The steps to process the fabrication of a metal line using deposition and polishing technology. (1) Fabrication of opening in mask structure on Si, (2) Trench formation by RIE, (3) Metal deposition step, (4) mechanical polish.

Filling the trenches becomes a significant challenge to the metal deposition community. Evaporation is not considered to be a manufacturable technique on the scale of today's integrated processing systems. Chemical Vapor Deposition technology (CVD) has been used successfully for the deposition of tungsten films, but is difficult to implement for Al or Cu metallurgies, which are the metals of choice for integrated circuit applications. In addition, they suffer from the inability to have high rate and to deposit alloys in a controlled way. Plating technology is a possibility for Cu-based metallization, but is not possible for Al based lines.

Therefore, sputtering has emerged as a promising candidate to fill these structures. However, conventional sputtering is inadequate to fill a high aspect ratio trench because of the high degree of sputtered atoms moving roughly parallel to the sample surface. Deposition under these circumstances leads to the build up of overhangs and eventual closure of the trench with a void formed inside. The problem of the laterally moving sputtered atoms can be alleviated by operating the system at very low pressure and interposing a directional filter between the cathode and the sample (5). The filter, known as a collimator, is simple a collector designed in the form of parallel tubes of moderate to high aspect ratio. (Fig. 6). Atoms moving in directions which are not roughly perpendicular to the film surface are deposited on the inner walls of the collimator. In effect, the collimator functions as a subtractive filter, removing the atoms which would lead to the formation of the overhangs. The net deposition rate in the bottom of the trench is not strongly affected, but the deposition rate on a planar surface can be strongly reduced (Fig. 7).

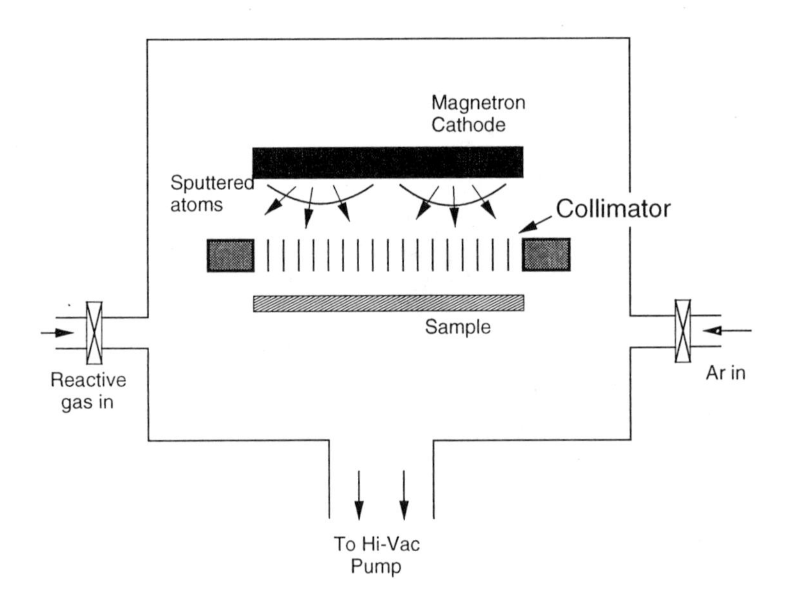

Figure 6. The configuration of collimators between the cathode and the sample.

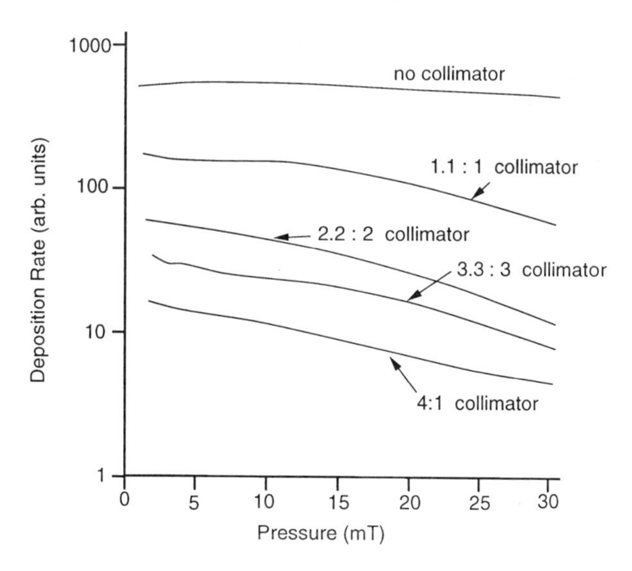

Figure 7. The effect of collimation on the deposition rate.

The deposition process, however, is sufficiently directional that it is possible to fill trenches with an aspect ratio of as high as 3.0. The principle advantage of the collimation process is that it has all of the intrinsic advantages of sputtering while requiring only minor equipment modifications. Most manufacturing scale systems can be outfitted for collimators without any significant modification to the chamber or related hardware.

A related issue for the sputtering of metal onto semiconductor wafers is the intrinsic non-uniformity of the magnetrons sputter deposition process. Etching takes place predominantly in the ExB track of the magnetron. The source of the sputtered metal tracks this path, also, and therefore leads to poor uniformity of deposition in virtually all geometries. The common fix for this problem is to move the sample in some form. However, for high-quality semiconductor processing, movement can often lead to particulate formation, which strongly reduces the yield of the process.

A solution to the intrinsic uniformity problem with the magnetron is to systematically move the etch track position during the deposition run. The most elegant means to do this is with a design known as the "rotating heart" magnetron. In this device, the etch track is set up by means of fixed, permanent magnets situated behind the cathode. The etch track is in the form of a heart-shape, with the point of the heart pointed to the edge of the cathode, and the two lobes of the heart near the center of the cathode (Fig. 8). The magnet assembly is then rotated about the center axis, and the heart shape sweeps around the cathode surface.

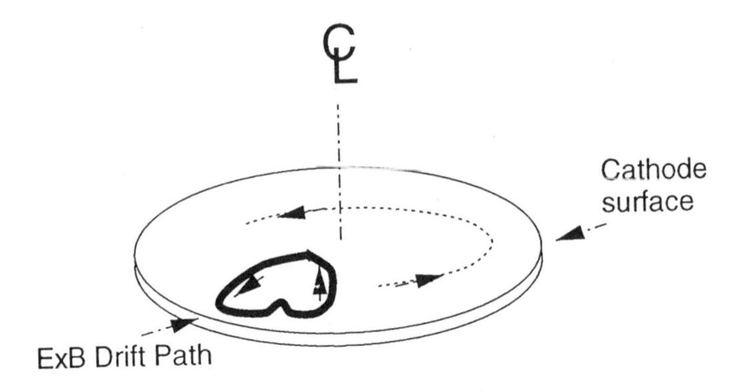

Figure 8. A heart shaped magnetron cathode design.

At any given instant in time, this design is obviously even less uniform than the conventional magnetron. However, after a complete rotation of the magnet assembly, the etched area can be made uniform to within a few percent. Subtle changes in the width or narrowness of the rotating heart shape can also influence the uniformity profile. The end result is a broad region across virtually the entire cathode surface which has been uniformly eroded, which leads to good uniformity of deposition across the sample surface.

UNBALANCED MAGNETRONS

Conventional magnetrons are actually designed to contain the plasma very close to the cathode surface for highest efficiency. The problem that then arises is that fairly low ion fluxes are then available to bombard a sample during deposition. This process, known as bias sputtering, has long been used to alter the composition, microstructure or planarization of the deposited film. Bias sputtering is used routinely in the field of rf-diode sputtering, but tends to fail with magnetron sputtering due to the too-good plasma confinement.

An obvious solution to this problem is to degrade the plasma confinement in the magnetrons and allow some fraction of the ions to leak out. The type of magnetron used for this is knowns as the "unbalanced magnetron" (6). The technique used is to strengthen one of the two poles (either center or edge) of a conventional magnetron. The stronger field can saturate the other pole, resulting in somewhat unconfined field lines which are no longer constrained to be located near the cathode (Fig. 9). Electrons from the cathode as well as the plasma are free to move along these field lines and can leave the vicinity of the cathode. This creates a weak potential gradient, which can cause ion diffusion away from the cathode and out to the sample region.

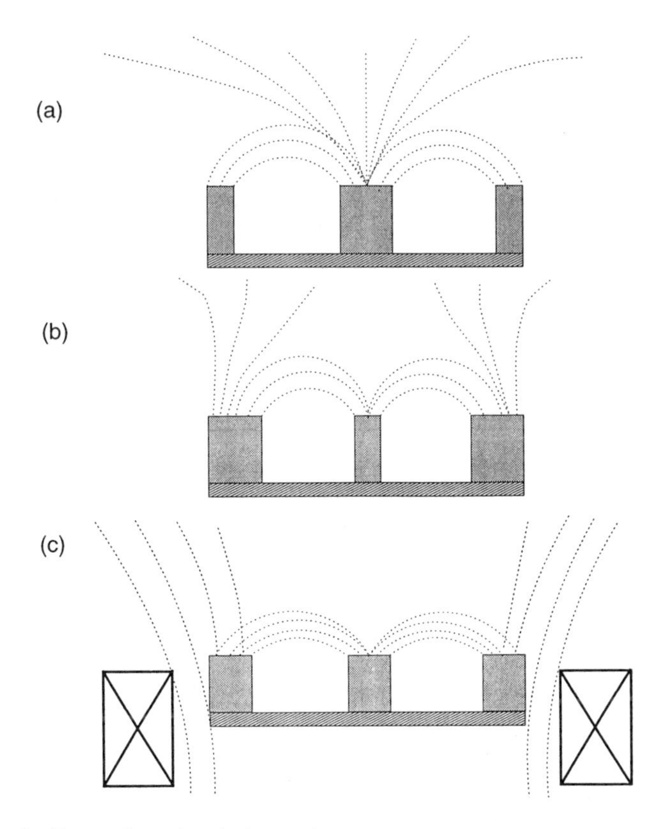

Figure 9. Examples of unbalanced magnetron designs.

The unbalanced magnetron is, in effect, a leaky magnetron. There is a slight increase in the discharge voltage, but it is not significant on the scale of the experiment. The ions which drift out from the cathode region can be accelerated to the sample surface in the form of a bias current. Large, unbalanced magnetrons have measures several milliamps per square centimeter bias currents at modest (-20) sample voltages with the sample in ion saturation current mode (7).

The primary application for unbalanced magnetron technology is in the formation of nitride films. This class of films tends to require either additional energy during deposition or else elevated sample temperature. Titanium nitride films, as well as other nitrides, tend to be hard, wear resistant and decorative. However, many of the substrates cannot be raised to the 700C or so temperature required to cause the nitride formation reaction. The solution has been large bias currents, which are now possible with the unbalanced magnetron. Several manufacturers have recently developed commercial deposition systems along this line, and the production of such things as nitride-coated drill bits have become commonplace.

CONCLUSION

Magnetron sputtering has long been used as a workhorse deposition tool in both the semiconductor industry, as well as numerous other industries, such as glass coating, the automobile industry, etc. Certain fundamental aspects of the sputtering process have limited its applicability in microelectronics applications as well as coatings. Recent developments of such techniques as the collimated sputtering system, the heart-shaped magnetron and the unbalanced magnetron have allowed significant additional capabilities to the field and have become in common usage in just a few years. Next generation tools, such as those which use post ionization of the sputtered flux in the form of a metal plasma, have the promise of additional capabilities which will be very useful in the development of higher density microelectronic circuits as well as other applications.

REFERENCES

1. S.M. Rossnagel and H.R. Kaufman, "Charge Transport in Magnetrons," J. Vacuum Science and Technology, A5 (1987) 2276-2279.
2. D. W. Hoffman, "A Sputtering Wind, J. Vacuum Science and Technology, A3 (1985) 561-565.
3. S. M. Rossnagel, "Gas Density Reduction Effects in Magnetrons," J. Vacuum Science and Technology, A6 (1988) 19-24.
4. S.M. Rossnagel, "Deposition and Redeposition in Magnetrons," J. Vacuum Science and Technology, A6 (1988) 3049-3054.
5. S.M. Rossnagel, D. Mikalsen, H. Kinoshita and J. J. Cuomo, "Collimated Magnetron Sputter Deposition", J. Vacuum Science and Technology, A9 (1991) 261-265.
6. B. Windows and N. Savvides, "Charged Particle Fluxes from Planar Magnetron Sputtering Sources," J. Vacuum Science and Technology, A4 (1986) 196-202.
7. W. D. Sproul, P. J. Rudnick, M. E. Graham and S. L. Rohde, Surface and Coatings Technology, 43/44 (1990) 270-275.

Section II

ATOMIC OXYGEN DURABILITY EVALUATION OF PROTECTED POLYMERS USING THERMAL ENERGY PLASMA SYSTEMS

Bruce A. Banks, Sharon K. Rutledge, and Kim K. de Groh
NASA Lewis Research Center
Cleveland, Ohio

Curtis R. Stidham
Sverdrup Technology, Inc.
Brook Park, Ohio

Linda Gebauer and Cynthia M. LaMoreaux
Cleveland State University
Cleveland, Ohio

Abstract

The durability evaluation of protected polymers intended for use in low Earth orbit (LEO) has necessitated the use of large-area, high-fluence, atomic oxygen exposure systems. Two thermal energy atomic oxygen exposure systems which are frequently used for such evaluations are radio frequency (RF) plasma ashers and electron cyclotron resonance plasma sources. Plasma source testing practices such as sample preparation, effective fluence prediction, atomic oxygen flux determination, erosion measurement, operational considerations, and erosion yield measurements are presented. Issues which influence the prediction of in-space durability based on ground laboratory thermal energy plasma system testing are also addressed.

Plasma Synthesis and Processing of Materials
Edited by Kamleshwar Upadhya
The Minerals, Metals & Materials Society ©1993

Introduction

Atomic oxygen in LEO can rapidly cause oxidation of organic materials which have been typically used for spacecraft construction (ref. 1). Atomic oxygen protective coatings have been developed for purposes of increasing the durability of oxidizable materials needed for the construction of LEO spacecraft (ref. 2). Evaluation of the durability of protected polymers intended for LEO application can be accomplished by in-space testing on spacecraft such as the Long Duration Exposure Facility (LDEF), or by means of ground laboratory atomic oxygen exposure facilities (refs. 3-4). Although in-space durability evaluations can produce accurate indications of in-space durability of protected polymers such as those being considered for the construction of Space Station Freedom, differences between the fixed orientation of investigatory spacecraft, such as the Long Duration Exposure Facility (LDEF), and functional spacecraft, which frequently have sweeping atomic oxygen arrival, can result in uncertainty concerning accurate prediction of materials performance in the functional spacecraft environment. Ground laboratory simulation tests can be accomplished by energetic directed atomic oxygen systems that simulate the 4.5 eV atomic oxygen arrival on spacecraft surfaces, or by thermal energy plasma systems that allow less than 1 eV atomic oxygen arrival (refs. 5-6). Although energetic directed atomic oxygen beam systems hold potential to more accurately simulate the mechanisms involved with LEO atomic oxygen attack, the practicality of using such systems has been frequently limited due to the expense associated with producing homogenous, large-area, high-fluence exposures. Low energy RF plasma and electron cyclotron resonance microwave plasmas have routinely been used to inexpensively evaluate the durability of protective materials intended for LEO application. The use of such low energy plasma systems does require care with respect to the operational practices and interpretation of results in order to obtain reliable predictions of in-space durability based on results of ground laboratory testing. The recommended operational practices for thermal energy plasma testing and considerations pertinent to in-space projection of durability for protected polymers is the subject of this paper.

Apparatus

Plasma Ashers

Radio frequency plasma ashers using a 13.56 MHz capacitively-coupled air or oxygen plasma have been used to generate high flux atomic oxygen in the laboratory. Figure 1 shows a photograph of a typical RF plasma asher system. Plasma ashers operated on air provide exposure to oxygen and nitrogen atoms, molecules, and excited species at pressures of 70-120 m torr, producing an oxygen-dominated reaction with polymer surfaces which is equivalent to an effective flux on the order of 10^{15} atoms/(cm^2-sec), based on erosion of polyimide Kapton H. Such effective fluxes produce erosion of unprotected Kapton exceeding that which would occur in LEO at 400 kilometers by an order of magnitude. The presence of nitrogen in such systems has been shown to have a negligible effect on the erosion processes of polymers (ref. 7). Samples exposed in RF plasma ashers are subjected to isotropic arrival of the plasma species. This typically results in minor roughening of the surface of protected materials. Although the surface is irregular, isotropic atomic oxygen arrival does not produce the high-aspect-ratio microscopic cones that are witnessed on all materials which have volatile oxidation products after exposure to directed ram atomic oxygen in space. Plasma ashers also produce intense 130 nm radiation, which may affect the erosion rate of some materials (ref. 6).

Electron Cyclotron Resonance Oxygen Sources

Atomic oxygen produced by electron cyclotron resonance plasmas from 2.45 GHz excitation can produce broad area directed or scattered isotropic atomic oxygen arrival. Figure 2 shows the NASA Lewis Research Center electron cyclotron resonance oxygen facility, configured in a mode of operation which allows an expanding atomic oxygen beam to be scattered off of fused silica surfaces to produce isotropic arrival of thermal energy (~0.04 eV) atomic oxygen. The samples exposed to atomic oxygen are located below a triangle of fused silica, which also has an aluminum foil strip within it. This triangle blocks the direct arrival of atomic oxygen and ionic oxygen, as well as the intense 130 nm radiation produced by the oxygen plasma, from arriving on the test sample surfaces. As can be seen in figure 2, the samples can also be exposed to controlled vacuum UV radiation by means of deuterium lamps located above and to either side of the 130 nm radiation blocking glass prism. Atomic oxygen can also scatter off the outer glass enclosure, thus reducing the population of ions and excited state species. Based on biased planar probe measurements in the sample plane, the fractional oxygen ion content is thought to be on the order of 10^{-3} to 10^{-4}. If one removes the glass fixturing, then directed thermal atomic oxygen arrival occurs, which produces conical surface microstructures identical to that observed in space. The directed beam has a fractional ion content with twice the ion current density as the scattered isotropic arrival configuration. The average oxygen ion energy is 13 eV. The 1000-watt microwave discharge uses an oxygen input flow of 35-50 standard cubic centimeters per minute. The facility pressure during source operation is maintained at 6 x 10^{-4} torr by means of a 10", 6100-liters-per-second diffusion pump, and a 115-liters-per-second blower, followed by a roughing pump, all operated with perfluorinated ether (Fomblin) oil.

Figure 1 - RF plasma asher.

Plasma Source Operational Practices

The previously described thermal energy plasma systems can be highly effective in evaluating the durability of protected polymers, providing that care is taken in the preparation, operation, and measurement of the samples placed within these facilities for evaluation. This section describes the techniques which have been found most reliable in obtaining useful information from such facilities.

Sample Preparation

Cleaning

The samples to be evaluated for atomic oxygen durability should be chemically representative of materials which would be used in space. Thus, the surface chemistry of the samples should not be altered by exposure to chemicals or cleaning solutions which would not be representatively used on the functional materials to be used in space. Wiping samples or washing them may significantly alter surface chemistry and atomic oxygen protection characteristics of materials, and is therefore not recommended. However, if the typical use in space will require solvent cleaning, then such cleaning should be performed to simulate actual surface conditions expected.

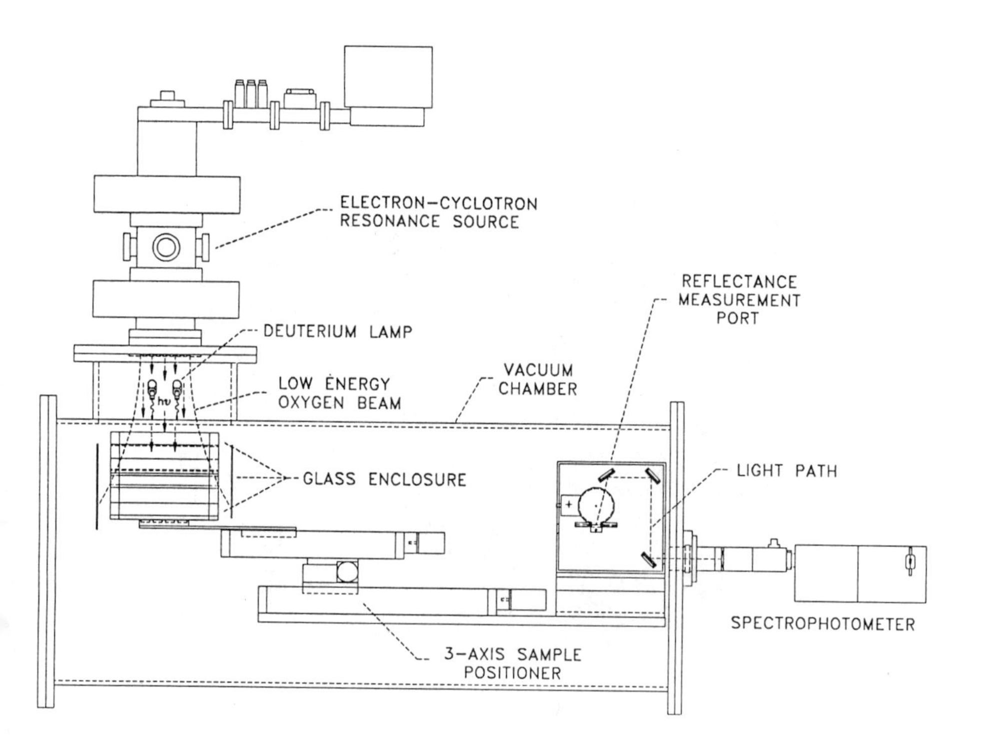

2a - Side view schematic.

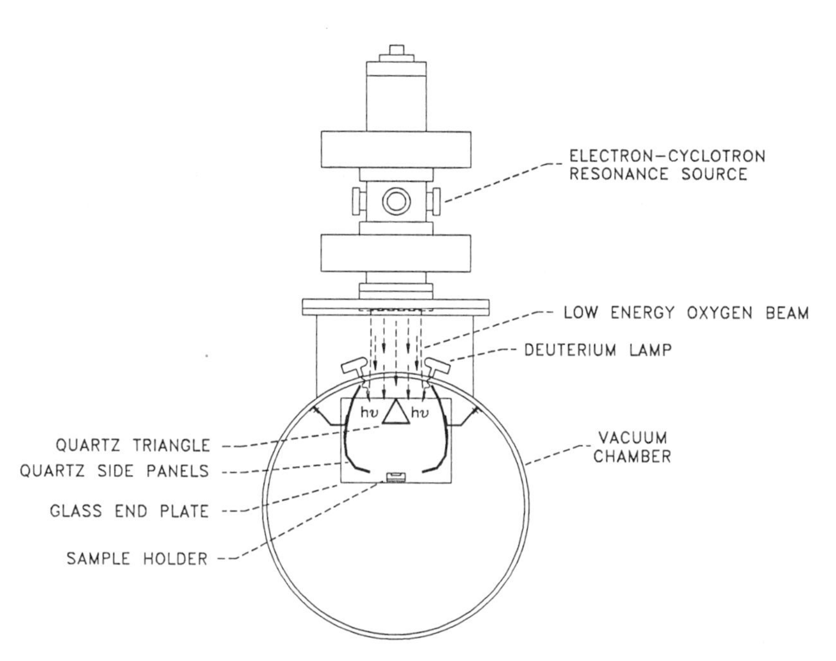

2b - End view photo and schematic.

Figure 2 - Electron cyclotron resonance atomic oxygen source.

Exposure Area Control

Masking. Frequently it is desirable to limit the exposure of atomic oxygen to one side of a material or a limited area on one side of the material. This can be done by wrapping a metal foil around the sample such as aluminum foil, or by placing a glass slide against one surface of the sample. If a metal foil is used to prevent atomic oxygen interaction on a portion of the sample, it is recommended that the foil be in intimate contact with the material to prevent partial exposure of the masked areas. Metal foil masking for exposure area control is acceptable for plasma systems in which the RF excitation region is separate from the sample exposure region. This minimizes electromagnetic interactions of the metal foil with the plasma that may cause sample heating and anomalous atomic oxygen fluxes to occur locally. Thus, metal foils are recommended for flowing afterglow and blow-by plasma configurations, and not recommended for use within the central RF discharge of a conventional plasma asher. The use of glass microsheet or slides to control the exposure area of samples must be limited to situations in which the samples are in intimate contact with the glass. This prevents excitation of any space which could occur between the sample and the glass, and provides assurance that areas which are to be protected, in fact have no atomic oxygen exposure.

Cladding. Samples which are coated with protective coatings on one side can be clad together by means of adhesives to allow the protective coating to be exposed on both sides of the sample. The use of thin polyester adhesives (or other non-silicones) is recommended to perform such cladding. The use of silicone adhesives should be avoided because of potential silicone contamination of the sample. Although cladding allows samples to be tested with the protective coatings on both faces, edge exposure of the samples and their adhesives does occur, and should be accounted for in calculating erosion characteristics of the desired surfaces.

Dehydration

Because most non-metals and non-ceramic materials contain significant fractional quantities of water, it is recommended that dehydrated thin polymer film samples be used to minimize confusion between oxidation and dehydration. Samples of a thickness less than 0.127 mm (0.005") should be dehydrated in a vacuum of a pressure less than 200 torr for a duration of at least 48 hours prior to sample weighing to ensure that the samples retain negligible absorbed water. Thicker samples should be dehydrated and weighed periodically until weight loss indicates that no further water is being lost. Multiple samples can be dehydrated in the same vacuum chamber, provided they do not cross contaminate each other, and that they are not of sufficient quantity so as to inhibit uniform dehydration of all the samples.

Weighing

Because hydration occurs quickly after removal of samples from vacuum, weighing the samples should occur within five minutes of removal from vacuum dehydration chambers or plasma ashers. Reduction of uncertainty associated with moisture uptake can be minimized by weighing the samples at measured intervals following removal from vacuum and back-extrapolating to the mass at time of removal from vacuum.

Handling

The atomic oxygen durability of materials with protective coatings may be significantly altered as a result of mechanical damage associated with handling. It is recommended that samples be handled in a manner which minimizes abrasion and flexure. The use of soft, fluoropolymer tweezers is recommended for handling polymeric films with protective coatings.

Effective Fluence Prediction

Witness Samples

It is recommended that samples of witness coupons be exposed to atomic oxygen simultaneous with test samples to enable calculation of the effective atomic oxygen exposure. If protective coatings on Kapton are being evaluated, it is recommended that Kapton witness coupons be used to enable the calculation of the effective atomic oxygen fluence. The evaluation of protective coatings on any substrate material should be accompanied by evaluation of that material without protection, provided in-space data exists concerning the erosion yield of that substrate material. If in-space erosion yield data for the substrate material does not exist, then it is recommended that Kapton H or Kapton HN be used as a witness material to calculate effective fluence based on an assumed in-space atomic oxygen erosion yield of 3.0 x 10^{-24} cm^3/atom. If high fluence exposure is necessary, polymeric sheets are usually too thin to survive long exposures. As a result, thick coupons of polyimide or graphite are suggested to be used for high fluence weight loss measurements. The atomic oxygen erosion yield of pyrolytic graphite relative to Kapton H is different in a plasma facility than in space. Unless both witness sample and test sample are of the same composition, and have known atomic oxygen erosion yields based on in-space testing, one should convert the mass loss of the pyrolytic graphite to the equivalent loss of polyimide Kapton H or HN. This can be accomplished by simultaneous plasma exposure of pyrolytic graphite and Kapton, and will enable the effective fluence to be calculated in terms of Kapton effective fluence, which is the accepted standard.

Test and Witness Sample Position and Orientation

Plasma facilities which have capacitively-coupled plasma excitation electrodes surrounding the sample test chamber usually have atomic oxygen axial and radial density gradients. In addition, the atomic oxygen flux on one side of the sample is usually different than that on the opposite side. Minimization of errors in effective atomic oxygen fluence will be achieved if witness samples are placed as close as possible to the same axial and radial locations as the test sample. The use of witness samples of the same size and shape as the test samples is also recommended. If similarity of position and orientation cannot be achieved, it is recommended that axial, radial, and orientation erosion rate characterization of the plasma chamber be performed to allow prediction of the atomic oxygen fluence at the site of the witness coupon.

Inspection Validation of Witness Sample Erosion

Kapton witness samples should be visibly inspected and compared with previously exposed witness samples which have demonstrated acceptable performance to validate that contamination of the surface of the sample has not occurred. Contamination can look like oil spots on the surface, a protective thin film, or other optical deviation from a normally diffusely reflecting exposed surface. The effective flux for the witness sample should also be compared with that from tests previously known to be acceptable, such as those performed in the same facility, to assure that neither contamination nor anomalous operation has occurred.

Witness Sample Weighing

Witness samples should be weighed within five minutes of removal from the vacuum chamber. Only one sample should be removed at a time for weighing. The rest should remain under vacuum. When witness samples are of the same chemistry as the substrate of the protected samples, it is important that both samples are weighed as close to the same interval of time after removal from vacuum. Accurate sample weight calculation, taking into account rehydration, can be achieved as described in the previously discussed section entitled "Weighing."

Atomic Oxygen Flux Determination

Relative effective fluences can be estimated from erosion of Kapton witness samples; however, measurement of absolute fluences is difficult because of scale factors associated with differences in erosion yields of thermal species in the plasma and in-space energetic species. The erosion yield, which is material-dependent, also appears to be substantially dependent upon oxygen atom energy. To assist in the prediction of in-space performance based on plasma testing, it is desirable to obtain both a direct measurement of the atomic oxygen fluence in the plasma facility, as well as an effective fluence for the exact material being tested by comparison with in-space performance. The use of pressure change measurements when the plasma is off, and then turned on, may allow an estimate of the atomic oxygen density in the plasma, provided correct assumptions are made with regard to the plasma temperature and the accommodation coefficient for the pressure sensing surfaces. Use of calorimetry to determine oxygen flux to a surface is an inexpensive and relatively simple method. However, the use of a thermocouple within the plasma causes excessive heating, leading to inaccurate temperature measurements. In addition, secondary reaction processes can occur on the calorimeter surface above certain temperatures. For these reasons, it is required that calorimetric probes be well-shielded from stray RF radiation, and be placed just outside the plasma.

Recession Measurement

Erosion yield measurements may be carried out by directly measuring the thickness of samples using cross-section photomicrographs. This method of determining erosion yield is independent of mass, as well as the density of the sample. A protected or shielded area must be available on each specimen during the actual exposure to establish an initial thickness or reference surface of the material. The area can be protected or shielded by a removable tape, a glass or metal solid shield, salt (NaCl) spray particles, or an electro-formed mesh held in intimate contact with a smooth test surface. The thickness loss is divided by the atomic oxygen effective fluence to obtain an erosion yield. This method avoids the uncertainties of mass loss due to outgassing and the need to determine exposure areas. Variations in the as-manufactured thickness are a potential source of uncertainty if a protected reference surface is not used. Uncertainty in the thickness measurement, as well as the magnification, must be considered in computing the overall uncertainty of the computed erosion yield. Recession measurements can be made by scanning electron microscopy or atomic force microscopy.

Plasma Facility and Operational Considerations

Contamination

Silicones present in an atomic oxygen plasma facility have been demonstrated to cause glassy films to deposit on witness coupons, as well as test coupons. Silicones may be present in the vacuum facility as a result of their presence in vacuum grease, vacuum seals, pump oil, or silicones in the samples themselves. Once silicones have been exposed to atomic oxygen, silicone contamination of the plasma chamber has probably occurred and removal of the source of the silicone contamination, such as a silicone sample, does not cause an immediate cessation of silicone contamination. It is therefore recommended that hydrocarbon-based vacuum greases and pump oils (when air is the source gas) or Fomblin (when oxygen is the source gas) be used in the plasma atomic oxygen facilities instead of silicones. Petroleum jelly has been found to be acceptable for use as a vacuum grease for plasma ashers. It is further desirable that the exposed area of the rubber vacuum seals be minimized to reduce degradation of the seals themselves, and to minimize reaction products being introduced into the plasma chamber.

Pressure

Minimization of backstreaming of roughing pump oil can be achieved if plasma facilities are operated at a sufficiently high pressure. Operation at pressures which allow backstreaming of roughing pump oil can result in sample or witness sample contamination due to pump oil oxidation products being deposited in the plasma facility. Typically, pressures of >60 m torr have been found to be necessary in plasma ashers to prevent backstreaming of roughing pump oil.

Plasma Chamber Degassing

To minimize outgassing contributions of vacuum chamber adsorbed gasses and moisture from contributing to the plasma species, it is recommended that room temperature chambers be evacuated to a pressure of less than 150 m torr for at least 30 minutes prior to initiation of the plasma.

Plasma Density and Uniformity

The visual intensity of the plasma glow has been found to be a sensitive measure of atomic oxygen flux. The operation of RF plasma ashers at conditions which produce run-to-run temporal uniformity, as well as spatial uniformity, is therefore recommended.

Continuous Versus Incremental Ashing

Some types of protective coatings on organic substrates develop mechanical stresses at defect sites in protective coatings where atomic oxygen undercutting has occurred. Exposure to air or humidity can cause mechanical stresses at the defect sites sufficiently high enough to propagate a tear in the protective coating in the vicinity of the defect. Thus, periodic exposure to air may cause accelerated degradation at defect sites, thus producing pessimistic results, whereas continuous high fluence exposure may produce results which are more representative of in-space functional exposure.

Oxidation Product Interactions

Samples exposed to atomic oxygen may have oxidation products which themselves enhance the reaction rate with either the sample itself or the witness sample. Oxidation product interaction problems are most easily detected by comparing witness sample effective fluence predictions with previous tests operated under the same conditions with only the witness sample present.

Temperature

The temperature of samples tested in plasma atomic oxygen exposure facilities should be operated as close as possible to their actual in-space functional temperature. Polyimide Kapton witness samples should be operated as close as possible to room temperature since their erosion yield was based on in-space measurements near room temperature. The use of samples which contain metals may significantly elevate the temperature of the sample if placed within the cavity of a capacitively-coupled plasma chamber, and therefore may anomalously increase the oxidation rate.

Plasma Composition

Typically, plasma atomic oxygen exposure is performed in either pure oxygen or air plasmas. Although other composition plasmas may produce significantly higher reaction rates, the lack of validity of reliable extrapolation to in-space results suggests that only oxygen or air plasmas should be used. Mechanistic oxidation information can be obtained through the use of isotope-enriched O^{18} oxygen to allow discrimination of native O^{16} oxides from those produced by atomic oxygen. For most polymeric materials, air plasmas have been found to be acceptable for exposure, and produce similar results to oxygen plasmas. The vacuum ultraviolet radiation content in the plasma may result in increased degradation of UV-sensitive materials such as Teflon (ref. 6). The presence of excited states in the plasma may also contribute to the difference in erosion yields of some materials with respect to each other when exposed in the plasma compared to space. However, materials which are unreactive in space, are unreactive in plasmas, and very reactive materials in space are found to be also very reactive in plasmas.

Erosion Yield Measurement

Test Sample Weighing

Test samples should be weighed at the same time interval and state of dehydration as the witness coupons as described in the previous sections entitled "Weighing," and "Witness Sample Weighing."

Profilimetry

Sample oxidation characterized by recession measurement may be possible if the sample material has gaseous oxidation products, its surface is smooth, and masking for exposure area control is executed with the mask in intimate contact with the sample. Because most polymeric materials do not have highly smooth surfaces, it is recommended that profilimetry be performed at multiple locations, and that high fluences be used such that the profilimetry step is large compared to the surface roughness.

Erosion Yield Calculation

The erosion yield of a material can be calculated based on mass loss, and is typically given in units of cubic centimeters per incident oxygen atom. The erosion yield is given by:

$$E_S = \frac{\Delta M_S}{A_S \rho_S F} \tag{1}$$

where:

ΔM_S = *mass loss of the sample, grams*

A_S = *surface area of sample exposed to plasma,* cm^2

ρ_S = *density of the sample,* $grams/cm^3$

F = *fluence,* $atoms/cm^2$

Because the fluence is typically based on the effective fluence of a witness coupon, we have the erosion yield:

$$E_S = \frac{\Delta M_S A_W \rho_W E_W}{\Delta M_W A_S \rho_S} \tag{2}$$

where:

A_W = *the area of the witness coupon, cm*^{2}

ρ_W = *the density of the witness coupon, grams/cm*^{3}

E_W = *in-space erosion yield of witness coupon, cm*^{3}*/atom*

ΔM_W = *mass loss of the witness coupon, grams*

Protective Coating Performance

Protective coating effectiveness, P, is a measure of how well a protective coating prevents oxidation of an underlying polymer. Protective coating effectiveness, a unitless number, is given by:

$$P = 1 - \frac{\Delta M_C A_U}{\Delta M_U A_C} \tag{3}$$

where:

ΔM_C = *mass loss of coated material, grams*

ΔM_U = *mass loss of uncoated material, grams*

A_C = *surface area of coated material, cm*^{2}

A_U = *surface area of uncoated material, cm*^{2}

Thus, a perfect atomic oxygen protective coating would have a protective coating effectiveness of 1, whereas a non-coated substrate would have a protective coating effectiveness of 0.

Relative reactivity, R, a unitless number defined as:

$$R = \frac{\Delta M_C A_U}{A_C \Delta M_U} = 1 - P \tag{4}$$

is another measure of coating performance where $0 \le R \le 1$.

Projection of in-space durability of a protected polymer based on ground laboratory plasma ashing requires correction of mass loss data to take into account the very high absolute atomic oxygen flux in ashers, knowing that the effective flux in an asher is much lower than the absolute flux. This correction is needed because undercutting protective coatings, both in space and in plasma ashers, is strongly dependent upon oxidation by thermally accommodated atoms.

Projection of In-Space Durability Based on Ground Laboratory Thermal Energy Plasma System Testing

The ability to predict in-space durability of protected polymers based on ground laboratory thermal energy plasma system testing is highly dependent upon the ability to quantifiably understand the reaction processes occurring in both space and the thermal energy plasma systems. Two parameters which greatly influence the outcome of such predictions are the probability of atomic oxygen reaction upon impact and the degree to which the atomic oxygen thermally accommodates upon impact. The reaction probability for energetic (approximately 4.5 eV) space ram atomic oxygen can be estimated by knowledge of the erosion yield of materials in space and assumptions concerning the oxidation product species. For example, for Kapton H polyimide, which has an in-space erosion yield of 3×10^{-24} cm^3/atom, initial impact reaction probabilities range from 0.0114 to 0.161, depending upon the oxidation product species. If one assumes an average between the extremes of oxygen possible in the oxidation products, then an average of 0.138 is predicted for the initial impact reaction probability (ref. 8). The reaction probability for room temperature thermal energy (approximately 0.04 eV) is thought to be substantially below the more energetic in-space values. Table I lists predicted atomic oxygen reaction probabilities in various environments based on the findings of several investigators. In Table I the reaction probabilities for both room temperature, thermally accommodated, and in-plasma atomic oxygen were calculated on the basis of assuming an in-space reaction probability of 0.138, and then adopting the various investigators' models for reaction probability dependence upon energy or environment. As can be seen from Table I, there is a span of three orders of magnitude variation in predicted reaction probabilities for room temperature thermally accommodated atomic oxygen. The reason that the reaction probability for room temperature atomic oxygen in a plasma is thought to be approximately four times that of ground state thermally accommodated atomic oxygen, is the presence of excited state species such as O^1D oxygen. The factor of four increase in reaction probability was determined as a result of Monte Carlo computational modeling to replicate observed atomic oxygen undercutting profiles at defect sites in protective coatings on Kapton. Based on the reaction probabilities indicated in Table I, thermal energy plasma systems require significantly higher actual fluxes than in-space to produce the same rates of oxidation. The consequences of this are illustrated in figure 3. As illustrated in figure 3, ground laboratory plasma systems are operated at high actual flux levels because of the low reaction probability of the low energy atomic oxygen. Thermal energy oxygen plasma systems produce an effective fluence computed by how much actual fluence in space would be necessary to cause the same amount of surface recession based on the in-space erosion yield of the particular material. Because far more actual atoms are necessary to impact a surface to produce the same surface recession as would occur in space, atomic oxygen undercutting at defect sites progresses much more rapidly in thermal energy plasma systems than in space, where erosion below a defect site in a protected polymer is dominated by the initial impact reaction. Depending upon the probability of thermal accommodation upon impact in space, the mass loss per unit area per fluence in space for protected polymers is thought to be between 10^{-2} to 10^{-4} that measured in thermal energy plasma systems based on effective fluence (ref. 4). It is likely that upon more detailed analysis of LDEF and EOIM-III data, as well as future in-space experiments, a more well-defined quantification of the correlation between ground laboratory and in-space erosion of protected polymers will occur.

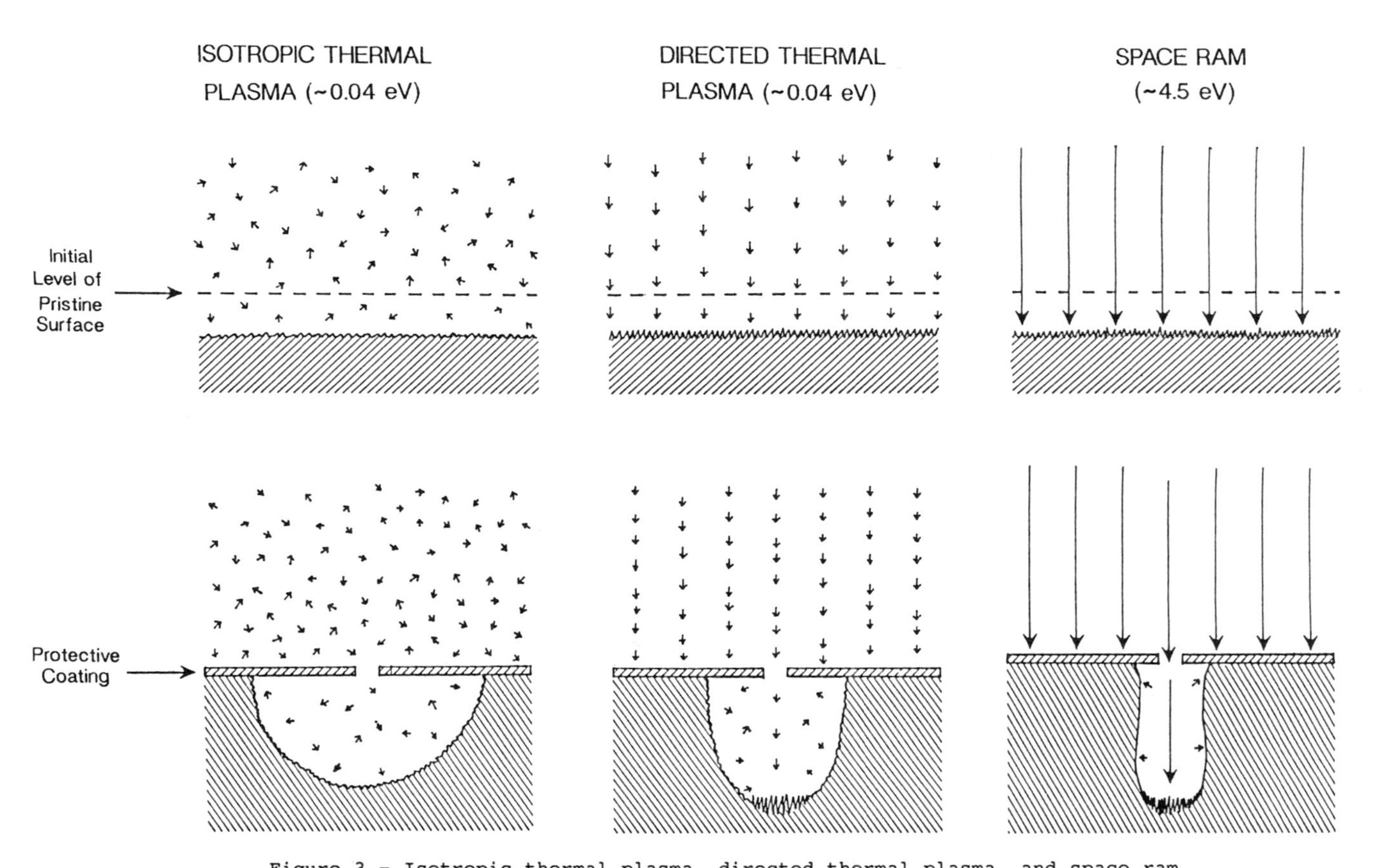

Figure 3 - Isotropic thermal plasma, directed thermal plasma, and space ram atomic oxygen interactions on uncoated polymers and at defect sites in protective coatings on polymers.

TABLE I. - ATOMIC OXYGEN REACTION PROBABILITY WITH POLYIMIDE KAPTON H

Environment	Energy, eV	Reaction Probability	Rationale for Prediction of Reaction Probability *	Ref
Space Ram (400 km, 996K, 28.5° inclination)	4.5 ± 0.8	0.138 ± 0.024	In-space erosion yield data and reaction product modeling	8
Room Temperature Thermally Accommodated (300K)	0.039	7.7×10^{-6}	Reaction probability proportional to $e^{-0.38/E}$	4,6
	0.039	1.0×10^{-5}	Reaction probability proportional to E^2	9
	0.039	2.1×10^{-3}	Monte Carlo modeling to match LDEF undercut profiles	10
	0.039	5.5×10^{-3}	Reaction probability proportional to $E^{0.68}$	11
Room Temperature Plasma (300K)	0.039	2.2×10^{-2}	Monte Carlo modeling to match undercut profiles	4,10

* E = atomic oxygen kinetic energy, eV

Summary

Thermal energy plasma systems such as RF plasma ashers and electron cyclotron resonance atomic oxygen sources can be of great assistance in determining the effectiveness of atomic oxygen protective coatings. The validity of the comparative information resulting from the use of such facilities can be greatly enhanced through the use of operational and testing practices which have been demonstrated to yield meaningful results. The projection of in-space durability based on ground laboratory thermal energy plasma system testing is highly dependent upon the reaction probabilities for energetic and thermally accommodated atomic oxygen. Although quantification of such numbers apparently varies over two orders of magnitude, analysis of LDEF, EOIM-III, and future flight experiments will more clearly quantify the correlation between erosion rates of protected polymers in thermal energy plasma systems with that expected in space.

Acknowledgements

The authors gratefully acknowledge the contributions of Bland Stein of NASA Langley, Matt McCargo of Lockheed Missiles and Space Company, and Gary Pippin of Boeing Aircraft Corporation for their contributions to the recommended practices for atomic oxygen exposure and analysis using plasma sources.

References

1. L.J. Leger and J.T. Visentine, "A Consideration of Atomic Oxygen Interactions with the Space Station," Journal of Spacecraft and Rockets, Vol. 23, No. 5, 1986, pp. 505-511.

2. B.A. Banks et al., "Ion Beam Sputter-Deposited Thin Film Coatings for the Protection of Spacecraft Polymers in Low Earth Orbit," NASA TM-87051, Proceedings of the 23rd Aerospace Sciences Meeting, Reno, Nevada, 14-17 January 1985.

3. B.A. Stein and H.G. Pippin, "Preliminary Findings of the LDEF Materials Special Investigation Group. LDEF - 69 Months in Space" (Paper presented at the First Post-Retrieval Symposium), NASA CP-3134, Part Two, 1991, pp. 617-641.

4. B.A. Banks et al., "The Use of Plasma Ashers and Monte Carlo Modeling for the Projection of Atomic Oxygen Durability of Protected Polymers in Low Earth Orbit" (Paper presented at the 17th Space Simulation Conference, Baltimore, Maryland, 9-12 November 1992).

5. B.A. Banks and S.K. Rutledge, "Low Earth Orbital Atomic Oxygen Simulation for Materials Durability Evaluation" (Paper presented at the Fourth European Space Symposium on Spacecraft Materials in a Space Environment, CERT, Toulouse, France, 6-9 September 1988).

6. S.R. Koontz, K. Albyn, and L.J. Leger, "Atomic Oxygen Testing with Thermal Atoms Systems: A Critical Evaluation," Journal of Spacecraft and Rockets, Vol. 28, No. 3, May-June, 1991.

7. S.K. Rutledge et al., "An Evaluation of Candidate Oxidation Resistant Materials for Space Applications in LEO," NASA TM-100122, 1986.

8. B.A. Banks et al., "Atomic Oxygen Undercutting of Defects on SiO_2 Protected Polyimide Solar Array Blankets" (Paper presented at the Materials Degradation in Low Earth Orbit Symposium of the 119th TMS Annual Meeting and Exhibit, Anaheim, California, 18-22 February 1990).

9. R. Krec, private communication with author, Physical Sciences, Inc., 27 October 1992.

10. B.A. Banks et al., "Atomic Oxygen Interaction at Defect Sites in Protective Coatings on Polymers Flown on LDEF" (Paper presented at the LDEF Materials Results for Spacecraft Applications Conference, Huntsville, Alabama, 27-28 October 1992).

11. D.C. Ferguson, "The Energy Dependence of Surface Morphology of Kapton Degradation Under Atomic Oxygen Bombardment" (Paper presented at the 13th Space Simulation Conference, Orlando, Florida, 8-11 October 1984).

LOW PRESSURE DEPOSITION OF DIAMOND FROM

WATER AND METHANOL GAS MIXTURES

Donald Gilbert, Richard Tellshow and Rajiv K. Singh

University of Florida
Department of Materials Science and Engineering
Gainesville, FL 32611

Abstract

We have used an electron cyclotron resonance enhanced plasma chemical vapor deposition system to deposit diamond thin films outside the active plasma region. The depositions were conducted at substrate temperatures of approximately 680 °C and at 100 mTorr of pressure using mixtures of water and methanol as the process gas. Substrates were positively biased, relative to ground, during deposition. In the absence of the positive electrical bias, no diamond growth was observed.

Plasma Synthesis and Processing of Materials
Edited by Kamleshwar Upadhya

Introduction

A great deal of interest was generated in industry when production of thin diamond films became possible by the chemical vapor deposition (CVD) method. This is due to diamond's many exceptional properties, such as high thermal conductivity, hardness, wide band gap and chemical inertness. Most CVD methods of diamond growth (such as hot filament and high pressure (>10 Torr) microwave, dc and rf plasma) may be termed "thermal" because they rely on thermal dissociation of gaseous precursor species.[1-4] Input gases such as CH_4, O_2, and H_2 are heated to temperatures greater than 2000 K to form active diamond-growth species. Under these conditions, electron and neutral gas temperatures are approximately the same. Almost any technique capable of heating a gas above 2200 K seems to be capable of depositing diamond thin films under the proper conditions of chemical composition and gas pressure.[1] The various techniques differ primarily in the type and intensity of secondary excitations which operate concurrently with thermal excitation of the gas mixture.

Using a technique called electron cyclotron resonance (ECR) enhanced CVD, low pressure (<100 mTorr), highly ionized plasmas can be generated. ECR occurs when microwave energy is resonantly coupled to electrons experiencing cyclotron motion in a magnetic field. This condition generates high energy electrons (>10 eV)[5], which then cause impact ionization and dissociation. True ECR does not occur above 10 mTorr due to the limited mean free path of electrons in the plasma. However, significant magnetic field confinement and enhancement of the plasma has been observed at pressures as high as 100 mTorr.[6] The generation of active growth species using this technique is achieved primarily by non-thermal mechanisms.

There are several possible advantages to the use of ECR plasmas for depositing diamond thin films. Because the temperature of the neutral gas and the chamber walls are nearly the same, and the pressures involved are in the molecular flow regime, it is relatively easy to generate a uniform distribution of activated species over a large area. Also, the low pressure of the plasma system may allow deposition to occur remote from the point of generation of the activated species. Finally, the low sensible temperature of the plasma does not contribute significantly to the heating of the substrate surface, making substrate temperature uniformity and control easier to maintain. Lower deposition temperatures are highly desirable to reduce thermal effects on substrates.

Magnetically enhanced microwave plasma enhanced CVD of diamond was first reported by A. Hiraki et al., who employed a magnetic mirror ECR design in which two electromagnets were used to couple a static B field to a microwave field.[6] The substrate was positioned within the plane of the ECR condition. Recently, Eddy et al. also reported the growth of diamond films at low pressures using CO and hydrogen gas mixtures. [7]

Researchers at the Research Triangle Institute have reported the deposition of diamond films from alcohol and water mixtures using a rf plasma system.[8] The use of H_2O instead of H_2 enabled a reduction in the substrate temperature necessary for diamond growth. The OH group is speculated to perform a similar function to atomic hydrogen, i.e. stabilization of the diamond phase and preferential etching of graphite. The use of H_2O instead of H_2 as a source gas also has many practical advantages, e.g. H_2O is cheaper and safer to handle than H_2.

Experimental Methods

Figure 1 shows a schematic diagram of the experimental set-up used in this research. Two rings of rare earth magnets (10 magnets per ring) are mounted on the exterior of a 15 cm inner-diameter by 15 cm long ECR module. The magnetic field at the pole face of each magnet is approximately 3 kGauss. Cusp-shaped zones of 875 Gauss field are generated inside the vacuum chamber. A turnstile coupler is used to direct a 2.45 GHz microwave field through a silica vacuum/microwave window at the top of the ECR module. At pressures below 150 mTorr, the coupling of the magnetic field to the electrons in the microwave plasma is strong enough to position the center of the plasma within the two rings of magnets. At pressures below 10 mTorr, electron cyclotron resonance occurs along cusp-shaped surfaces where the magnetic field strength equals 875 Gauss. A secondary ring of magnets is positioned below the ECR module for plasma confinement. The source gas mixture used was methanol and water. Methanol was introduced through a ring positioned below the base of the ECR module, while water was injected from the top. Substrates were placed on a heated platen which was electrically isolated to enable dc biasing. The mounting system allowed for variation of substrate height in the chamber, and depositions took place 3 to 5 cm below the ECR zone. A positive bias of 50 volts or greater, relative to ground, seemed to be necessary for growing diamond films. The substrates used for growth in these experiments were molybdenum plates scratched with submicron diamond powder.

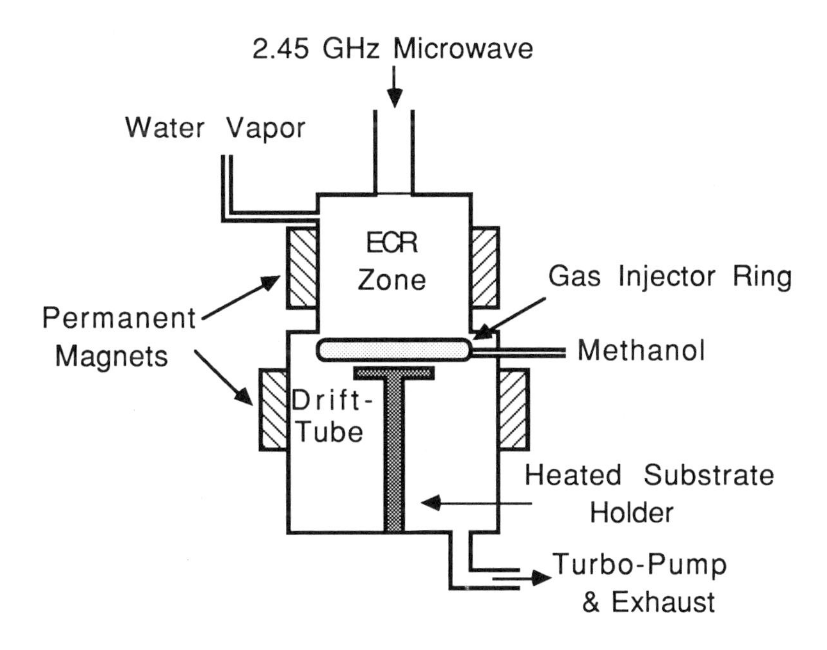

Figure 1. Schematic diagram of ECR-CVD system.

In order to characterize the plasma conditions, the electric current generated through the substrate as a result of the bias voltage was recorded. These measurements were made over a range of conditions, from 400 to 900 Watts of microwave power and pressures greater than or equal to 15 mTorr. Current readings were also taken at various substrate positions.

Results

Films were characterized using Raman spectroscopy for phase identification. Scanning electron microscopy was used to determine the film morphology.

Figure 2 shows a Raman spectrum of a film grown at a pressure of 100 mTorr and a substrate temperature of approximately 680 °C. The characteristic diamond peak at 1332 cm^{-1} is clearly visible. The broad background spectrum is indicative of a highly defective film containing non-diamond phase carbon. This film was deposited with a methanol/water volume flow ratio of 14:1 and microwave power of 900 W.

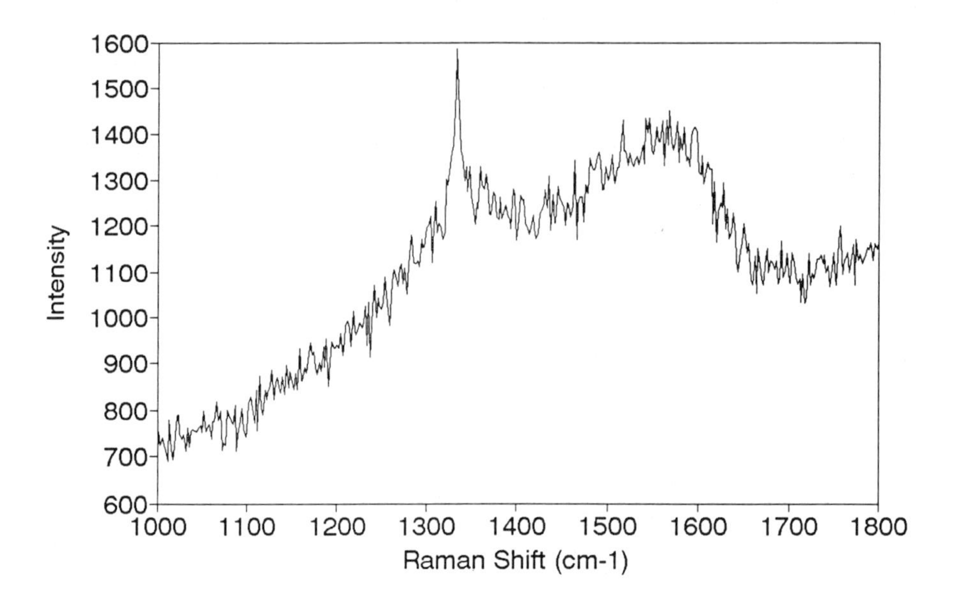

Figure 2. Raman spectrum of film showing diamond peak at 1332 cm^{-1}.

The surface morphology of the same film is shown in Figure 3. Figure 3 (a) shows a large coverage of particles, which are believed to be the result of residual diamond powder left on the surface for nucleation purposes from substrate preparation. A closer view of the central region is given in Figure 3 (b). The larger particles clearly show planar surfaces, which indicate growth of the nucleation material. The surrounding continuous film appears irregular in form. This irregular film is expected to be the source of the non-diamond component evident in Figure 2.

Figure 3(a). SEM of diamond film at 2000x showing general surface morphology.

Figure 3(b). SEM of diamond film at 20000x differentiating particles and continuous film.

A film grown under similar conditions is shown in Figure 4. The methanol/water volume flow ratio for this deposition was 18:1. Particles show evidence of faceting which would be consistent with growth. The same irregular continuous film is seen surrounding the larger particles.

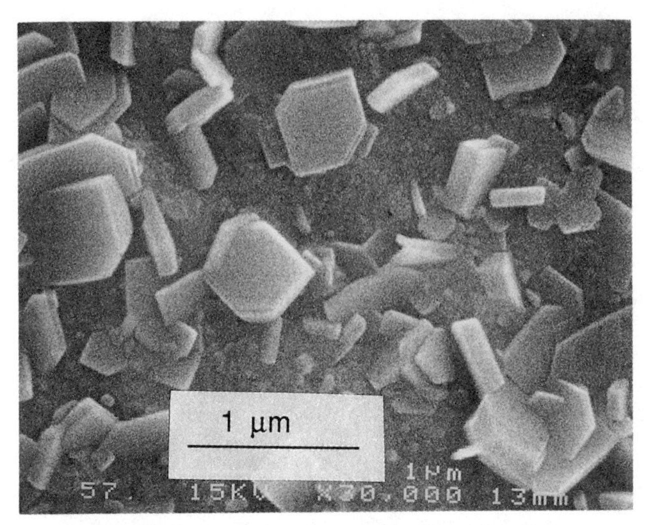

Figure 4. SEM of diamond film showing apparent faceting of particles.

The results of the plasma characterization measurements are shown in Figure 5. In Figure 5(a), which plots bias current against substrate position for an incident microwave power of 900 Watts, a peak in the bias current is seen for a substrate position (relative to the mid-point of the ECR module) of 7 cm. This corresponds to the substrate being positioned at the base of the ECR module. Figure 5(b) shows the bias-current behavior versus incident microwave power for different substrate positions at a pressure of 60 mTorr. This shows that, in general, the bias current is highest at a position 7 cm below the ECR zone. In the absence of a plasma, all current readings went to zero. It should be noted that the depositions characterized in Figures 2 - 4 took place roughly 3 cm below the midpoint of the ECR zone.

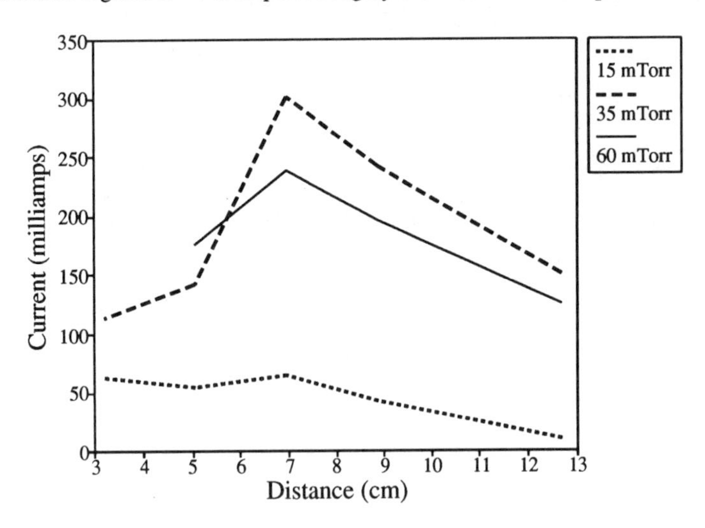

Figure 5(a). Plot of bias current vs. substrate distance from the midpoint of the ECR module.

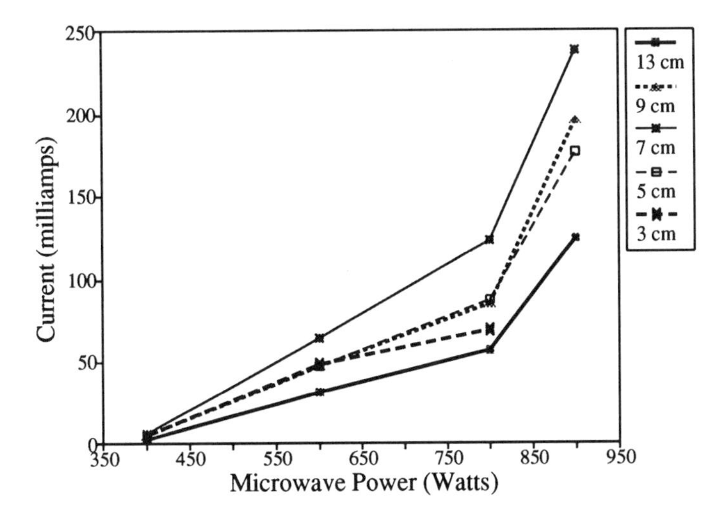

Figure 5(b). Plot of bias current vs. microwave power for various substrate positions.

References

1. P.K. Bachmann, D. Leers and H. Lydtin, Diamond and Related Materials 1, 1 (1991)

2. J. A. Thorton, in Deposition Technologies for Films and Coatings, ed. R. F. Bunshah (Park Ridge, NJ: Noyes Publishing, 1982), 23.

3. L. S. Plano, S. A. Stevenson and J. R. Carruthers, in Diamond Materials, ed. A. J. Purdes, et al. (Pennington, NJ: The Electrochemical Society, 1991 Vol. 91-8), 290

4. J. Wei, H. Kawarada, J. Suzuki, K. Yanagihara, K. Numata and A. Hiraki, in Diamond and Diamond-like Films, ed. S. Dismukes (Pennington, NJ: Electrochemical Society, 1989 vol 89-12), 393

5. Jes Asmussen, Journal of Vacuum Science and Technology A 7 (3) (1989), 883-892

6. A. Hiraki, H. Kawarada, J. Wei, J. S. Ma and J. Suzuki, in Diamond Optics III, SPIE Proc. Vol. 1325, ed. A. Feldman and S. Holly (Bellingham, WA: SPIE, 1990), 74

7. C. R. Eddy, D. Youchison and B. D. Sartwell, Surface and Coatings Technology, 48 (1991), 69

8. T. A. Rudder, G. C. Hudson, J. B. Posthill, R. E. Thonas, R. C. Hendry, D. P. Malta, R. J. Markunas, T. P. Humpreys and R. J. Nemanich, Applied Physics Letters, 60 (3) (1992)

PLASMA ACTIVATED SINTERING - A NOVEL, VERSATILE CONSOLIDATION PROCESS

Joanna R. Groza, Subhash H. Risbud and Kazuo Yamazaki

Department of Mechanical, Aeronautical and Materials Engineering,
University of California, Davis,
Davis, CA 95616-5294

Abstract

Plasma Activated Sintering (PAS) is a novel non-conventional consolidation process that combines plasma generation with resistance heating and pressure application. The plasma environment results in particle surface activation that enhances particle sinterability and reduces high temperature exposure. Examples of short-time densification of a large variety of materials such as metals, intermetallic compound and difficult to sinter ceramic materials are presented. The current understanding of plasma- and pressure-assisted sintering was used to identify the densification mechanisms for the PAS process.

Plasma Synthesis and Processing of Materials
Edited by Kamleshwar Upadhya

Introduction

Considerable attention has been recently centered on new materials that exhibit unusual properties such as superplasticity in normally brittle materials, ferromagnetic properties in non-ferromagnetic materials or electronic materials with controlled band gaps. Most of these materials are in powder form and attribute their unique properties to a metastable condition e. g. amorphous, nanocrystalline or nonequilibrium phase structures. Retention of this metastable condition is a major challenge for the consolidation step that involves high temperature exposure for usually long time. Plasma discharge is known to activate particle surfaces and thus enhance particle sinterability and reduce the high temperature exposure [1]. Additionally, pressure application assists the densification process by enhancing sintering processes and thus further reducing the high temperature exposure of the consolidating powders. The focus of the present paper is on a new, non-conventional plasma-assisted powder consolidation process, plasma activated sintering (PAS), that is capable of minimizing the thermal exposure of highly metastable materials. An additional pressure component further enhances the powder consolidation process. The (PAS) process is described and current results on the consolidation of various powder materials are presented. Next, based on the current understanding of plasma- and pressure-assisted sintering processes, we seek to identify the mechanisms by which this non-conventional processing method leads to powder compaction.

Experimental Procedure

Plasma Activated Sintering Process

The main components of the PAS consolidation process consist of plasma generation, resistance heating and pressure application. A schematic of the PAS process is shown in Figure 1. In this process, the loose powder is loaded in the punch and die unit and a biaxial pressure is applied for cold compaction. An instantaneous pulsed electric power is applied to create the plasma environment and thus activate the particle surface (control switch at right). The powder particles are subsequently resistance heated (control switch at left) while the uniaxial pressure is still applied to the sample in the sintering mold. The total pressure application time and resistance heating time are short but sufficient to cause plastic yielding and material flow for rapid consolidation.

Materials and PAS Consolidation Parameters

PAS consolidation has been applied to metallic (Ni), intermetallic compound (NiAl and Nb_3Al) and ceramic (AlN) materials. The metallic powder was commercially pure Ni. The NiAl powder has an average particle size of 6 μm and was supplied by XForm Inc. The PAS consolidation of Nb_3Al intermetallic started with elemental powders. Aluminum powder (purity 99.5% and particle size less than 44 μm) from Johnson Matthey and niobium powder (purity 99.9% and particle size less than 420 μm) from Aldrich Chemical were mechanically alloyed for 4 hours in a Spex Mixer/Mill 8000. The powder composition was 84.6 at.% Nb and 15.4 at.% Al. Details on the mechanical alloying process are given elsewhere [2]. The initial structure of mechanically alloyed powders consisted of an extended supersaturated Nb-based solid solution. Differential thermal analysis of this powder indicated the formation of the Nb_3Al intermetallic compound at 1150 K [2].

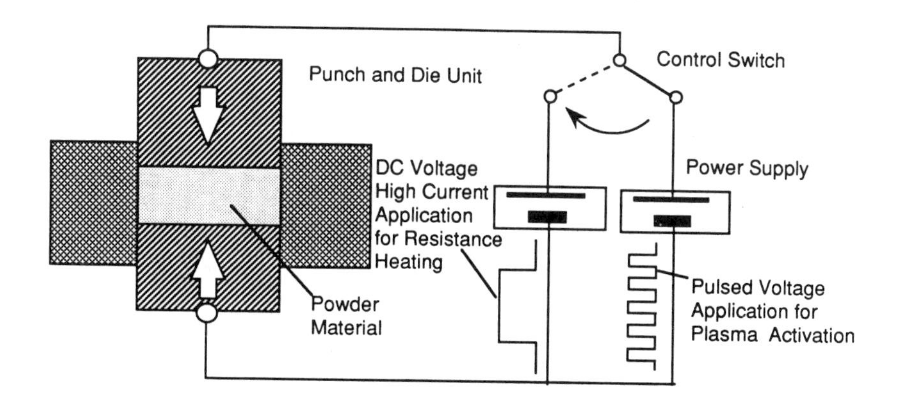

Figure 1. Schematic of PAS process.

The AlN powder was a developmental formulation (grade F) supplied by Tokuyama Soda, Japan. This powder has less than 10 ppm iron and a low oxygen content (1.0 %). The specific surface area is 3-4 m^2/g. The agglomerated size of initial powder is approximately 1.8 μm with a crystallite size of about 0.44 μm.

The PAS consolidation parameters of above powders are given in Table 1. A schematic of PAS stages and process parameters is shown in Figure 2.

Table 1. PAS Consolidation Parameters

Material	Pressure, MPa		Temperature, K	Time, min.	Environment
	Step 1	Step 2			
Ni	30	30	1173	~2	air
NiAl	12.5	40	977	4.75	vacuum
Nb-Al	-	30	1473	4	vacuum
AlN*	-	50	2000	5	air

*No additive used.

The characterization of sintered samples was achieved by density measurements using Archimedes' method, optical and transmission electron microscopy (Philips 400 T). A preliminary assessment of mechanical properties involved Vickers microhardness measurements under 1 N force.

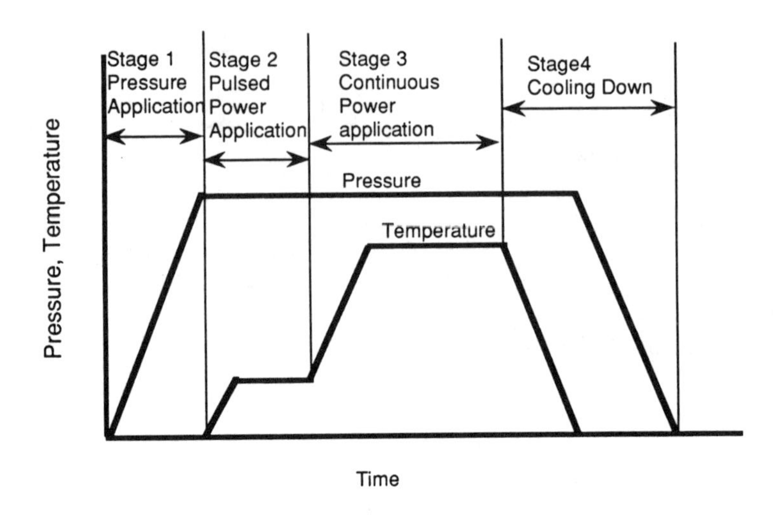

Figure 2. Schematic of PAS stages.

Results and Discussion

The final consolidated specimens were discs of 20 mm diameter and different heights (about 2 to 8 mm). The measured density values and relative densities of the consolidated specimens are given in Table 2.

Table 2. Densities and Hardness Values of PAS Consolidated Specimens.

Material	Density, g/cm^3	Relative Density, %	Hardness, MPa
Ni	8.33	93.6	-
NiAl	5.73	97.2	-
Nb-Al	7.62	~100	10000
AlN	3.24	99.3	8180

We mention that the density of pure nickel powder is that obtained in some preliminary tests that were performed in various consolidation conditions to study the PAS mechanism. The density of consolidated Nb-Al powders was very close to the theoretical density calculated by assuming equilibrium structure at the chemical composition of the studied alloy. The X-ray diffraction pattern of the consolidated specimen indicated the formation of the Nb_3Al compound with A15 structure [2]. This result is in agreement with previous work that showed that mechanically alloyed Nb-Al powders transform to the equilibrium A 15 phase after thermal annealing at 1098 K for 2 hours [3]. The TEM microstructure of consolidated Nb_3Al material is shown in Figure 3. The microstructure consists of equiaxed one-phase grains of about 200-300 μm and some smaller, probably two-phase grains. The phase identification of this material is in progress. No pores are observed in the microstructure and this correlates well with the high measured density values (Table 2). In the present mechanically alloyed powders that had an extended solid solution structure, the very short time at 1470 K during PAS consolidation was sufficient to promote Nb_3Al formation. Plasma environment may also be responsible for this rapid reaction rate between components to form the Nb_3Al compound.

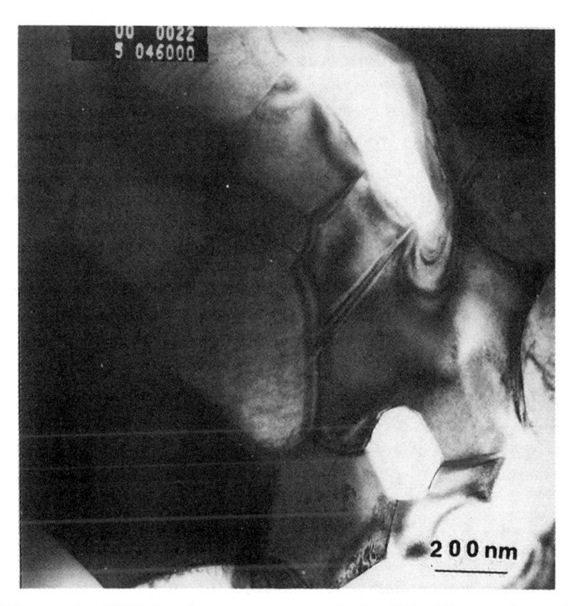

Figure 3. TEM microstructure of the Nb-Al powders that were mechanically alloyed for 4 hours and PAS consolidated.

Again, the densities of NiAl intermetallic are only preliminary values. However, these densities compare well with those achieved in NiAl by sintering at 1673 K and hot isostatic pressing (HIP) at higher temperatures (between 1523 and 1673 K) for significantly longer times (2 hours) [4].

The densities of covalent AlN ceramic range between 3.18 and 3.24 Mg/dm^3 (97.5 and 99.3 % of theoretical density, respectively). These densities are the highest achieved in non-doped AlN ceramic powders at sintering temperatures below 2023 K. The TEM examination of the consolidated AlN sample revealed equiaxed and uniform grains of about 0.77 μm [1]. Similar to Nb_3Al, for AlN structure, the lack of pores in the TEM observation correlates well with the high measured densities. The observed microstructure indicates that the grains in the consolidated material remained submicron size which is evidence for minimal grain growth during the PAS densification process. The measured hardness values for the consolidated AlN specimens is 8188 MPa.

Mechanisms of Densification

The mechanism of the accelerated PAS consolidation has not been studied yet. It appears to be a combination of the effects of electrical discharge, resistance heating and pressure application. The electrical discharge that creates plasma environment causes activation of particle surface that results in concentrated heat effects and hence, possible removal of surface oxides and entrapped gases and high heating rates. As shown by Upadhya [5], in the plasma generation step, numerous ions, electrons and energized atoms are created. The powder particles are exposed to bombardment of these highly activated species thus causing surface effects. These effects involve particle heating, surface cleaning and sputtering of loosely bonded atoms such those of entrapped gases. Consequently, new clean areas are exposed and an efficient bonding of powder particle surfaces may be developed during subsequent powders consolidation stages. The concentrated heat effects at particle surfaces may cause surface melting and oxide breakdown, similar to surface effects in electrodischarge machining [6]. In some other applications of pulsed electrical current such as for thermosynthesis of chemical compounds, an enhanced reaction was also observed [7]. This intensification of the reaction was attributed to the breakage of the oxide films on the surface of reacting powders. As shown above, such an accelerated reaction was noticed to form Nb_3Al compound during the PAS consolidation of Nb-Al powders. The intimate mixing of components by prior mechanical alloying may also contribute to this accelerated reaction. Based on the above information, we conclude that the same plasma effects may take place in the PAS process. The initial electrical pulse application breaks up the oxide films and exposes new clean areas for subsequent welding. This process may be partially regarded as an activation of particle surface or an "in-situ" cleaning action of powder particles by the removal of surface oxides and entrapped gases. This surface cleaning from oxides is important for two main reasons. Firstly, the newly cleaned and activated surfaces contribute to enhanced diffusion and welding during subsequent densification. Secondly, for many materials the oxide film on particle boundaries in consolidated material may have deleterious effects on their physical and mechanical properties. This in-situ cleaning action becomes more important for oxygen sensitive powders for which protection from oxygen contamination in early stages of powder preparation is practically impossible.

Surface melting may also be the result of resistance heating due to concentrated heat effects at powder particle surface. Generally, resistance heating is known to cause full densification by a liquid sintering mechanism [8]. Local surface melting occurs due to higher electrical resistance of surface oxides. In the plasma activated sintering process, though, the prior plasma activation step removes surface oxides and melting related to higher electrical resistance of surface films is less likely to occur. However, as already shown, surface melting may be caused by electrical discharge action. In our present TEM studies, no grain boundary phase has been observed to substantiate the liquid sintering mechanism. Characterization studies on various PAS consolidated materials with the aim of evaluating the grain boundary phase are in progress. Even for simple resistance sintering process, the microstructural characteristics shown to support the liquid formation are not convincing. For example, Liu and Kao reported consolidation of titanium by resistance sintering [8]. The optical microstructure of sintered titanium contains a "secondary" α that, according to them, has formed from a liquid phase. Unless our measurements are in serious error, the secondary α spacing is significantly different from initial powder size. If melting occurs at particle surfaces, we expect that secondary α will have the same distribution as initial particles.

Finally, mechanical pressure is being applied on heated particles. The applied external pressure assists the already started sintering process by efficient particle movement and pore closure. There are three main mechanisms that contribute to densification under pressure: plastic yielding, power-law creep and diffusional densification. While the latter mechanism exists in the pressureless sintering process and it is only enhanced by pressure application, the former two mechanisms are specific to pressure assisted densification. To identify each mechanism contribution, a short discussion on PAS densification conditions is necessary. The addition of external pressure facilitates full compaction during PAS process. This conclusion is supported by significantly shorter consolidation times in PAS than in hot isostatic pressing (HIP) while the level of the mechanical pressure is comparable. These shorter times are a good indication that solid state time dependent mechanisms (such as creep in HIP) do not contribute significantly to densification. Instead, pressure contribution may be mostly explained by considering the densification by plastic yielding. This densification, D_{yield}, is instantaneous in the initial stage of sintering and the starting density is given by the following equation [9]:

$$D_{yield} = \left(\frac{(1-D_o)\,P}{1.3\sigma_y} + D_o^3 \right)^{\frac{1}{3}} \tag{1}$$

where D_o is the initial relative density (usually 64%), P is the applied pressure and σ_y is the yield stress. The compact enters the final densification stage during plastic yielding only if pressure is high enough to cause yielding of the spherical shell surrounding each pore. Densification by yielding in the final sintering stage is given by [9]:

$$D_{yield} = 1 - \exp\left(-\frac{3}{2}\,\frac{P}{\sigma_y} \right) \tag{2}$$

However, the pressure holding time in the order of minutes at high temperature during PAS consolidation process may be considered as an indication that time dependent processes (such as power-law creep) may also play a role. When yielding stops, the contact area continues to deform by power-law creep. The densification rate by power-law creep [9] in the final stage when a pressure P is applied is:

$$\dot{D} = \frac{3}{2}\left(\frac{\dot{\varepsilon}_o}{\sigma_o^n}\right) \frac{D(1-D)}{\left[1-(1-D)^{\frac{1}{n}}\right]^n} \left(\frac{3}{2n}P\right)^n \tag{3}$$

where $\dot{\varepsilon}_o$, σ_o and n are material creep parameters. As already shown, power-law creep is expected to have little contribution to densification in PAS process in which the time interval at high temperature is characteristically short.

Densification by diffusion

Both grain boundary and lattice diffusion contribute to the matter transport from the contact zones to the surface of a sintering neck during the sintering process. Diffusion is accelerated by the externally applied pressure due to the increase in the surface-tension driving force and to the possible shear stress components on the grain boundary between particles that promote atom displacement. Naumov et al. have shown that in the plastic flow process under externally applied load, diffusional processes are extremely intensified due to a significant increase in lattice defect density [10]. In these conditions, they showed that the density of lattice defects increases to the level of premelting temperatures. The densification rate in the initial stage combines the diffusion kinetics for one contact with the increase in coordination during densification. Starting with the detailed treatment of Wilkinson and lately simplified by Ashby [11] this densification rate may be written as:

$$\dot{D} = A\frac{P_{eff}\Omega}{R^3}\left[\frac{D_v}{kT} R\ (D-D_o) + \frac{\delta D_b}{kT}\right] \quad (4)$$

where $A = c_1 (1-D_0)^2 / (D-D_0)^2$, c_1 is a numerical constant, Ω is the volume of the diffusing atom or molecule, R is the particle radius, D_v the lattice diffusion coefficient, δD_b the boundary diffusion coefficient times its thickness and kT has the usual meaning. Since many small contacts densify faster than a few large ones, the rate of the diffusional mechanism is strongly dependent on particle size R. This is in contrast to power-law creep densification rate which is independent of R (eq. 3). The diffusional contribution to the densification rate in the final stage is given by the following equation [11]:

$$\dot{D} = B\frac{P\Omega}{R^3}\left[\frac{D_v}{kT}R(1-D)^{\frac{1}{3}} + \frac{\delta DB}{kT}\right] \quad (5)$$

where $B = c_2 (1-D)^{1/2}$, c_2 is a numerical constant and the other symbols have been already defined. The predominant process varies with temperature. At high temperatures, lattice diffusion controls the rate, at lower temperatures, grain boundary diffusion takes over. The intensification of diffusional processes due to plastic flow is expected to occur in the final stage of densification when contact surface areas become important.
Initial plasma activation step contributes to enhanced diffusional densification, as well. As discussed by Johnson, the enhanced densification rate in rapid sintering processes originates from the suppression of surface diffusion [12]. The ratio of surface, J_s, to volume diffusion, J_v, is:

$$\frac{J_s}{J_v} = \frac{\Omega^{\frac{1}{3}} D_s}{2RD_v}\left(\frac{x^2}{rA}\right) \quad (6)$$

where D_s is the surface diffusion coefficient, x is the neck radius, r is the fillet radius and A is the neck surface area. The geometric term in parenthesis is large in the initial sintering stage and decreases with sintering time. Furthermore, surface diffusion predominates in the early stages of sintering since the activation energy for surface diffusion is less than for volume diffusion at lower temperatures. However, the PAS process appears to have the capability to lower the activation energy for volume diffusion by rapidly reaching high temperatures and thus enhance the densification rate. The rapid heating rates in PAS are achieved due to the combination of plasma generation and resistance heating. More importantly, high heating rates through the low temperature region are favorable for hindering surface diffusion process. It is known that densification is mostly controlled by volume diffusion while coarsening is due to surface diffusion. Therefore, by suppressing surface diffusion, grain growth process is minimized and retention of the initial fine grain structure becomes possible.

Pressure application results in enhanced densification but limited grain growth, as well. Generally, pressure is known to increase the driving force for densification with no effects on atomic mobilities so that grain growth is restricted [13]. The combined effects of plasma and pressure application in PAS contribute to enhanced grain boundary and lattice diffusion before surface diffusion can coarsen the microstructure and lower the driving force for densification. As a result, the consolidation time is very short thus preventing the undesirable grain growth process. Further evaluation is underway to account for the contribution of the individual sintering mechanism to the overall PAS consolidation process and will be reported later.

In conclusion, plasma activation and pressure application in the PAS process contribute to short consolidation time, high densities and minimal grain size coarsening. The enhanced sintering kinetics due to high-temperature plasma and pressure application are the principal factors contributing to full densification in the PAS process. The enhanced densification process contributes to the retention of the initial fine grain size in the undoped AlN and rapid component reaction for Nb_3Al intermetallic compound synthesis.

Conclusions

1. High densities compacts of various materials -metals, intermetallic compounds and covalent ceramics - were obtained by PAS consolidation.
2. The activation of particle surface and cleaning power of the plasma discharge confer unique densification ability for difficult to sinter materials such as covalent ceramics or highly oxygen sensitive powders.
3. The initial submicron grain size has been retained in some PAS consolidated specimens (AlN). This retention of grain size may be attributed to a significantly short high temperature exposure of powder particles.
4. The factors that contribute to a short time consolidation during PAS process are plasma activation, rapid resistance heating and mechanical pressure application.
5. It appears that the PAS capability of preventing coarsening resides in the suppression of surface diffusion and enhancing the lattice diffusion due to rapid heating.

Acknowledgements

The authors are thankful to Sodick Co., Ltd. in Japan for some PAS densification work and to Dr. G. Watson, NASA-Lewis Reserach Center for supplying NiAl powder.

References

1. J. R. Groza, S. H. Risbud and K. Yamazaki, J. Mater. Res. 1992, 7, 2643-2645.
2. J. R. Groza, Mechanical Alloying of Nb_3Al Compound, to be published.
3. R. C. Benn, P. K. Mirchandani and A. S. Watwe, in Modern Developments in Powder Metallurgy, APMI, MPIF, Princeton, NJ, 21, (1988) p. 479-93.
4. J. C. Murray, R. Laag, W. A. Kayser and G. Petzow, in Advances in Powder Metallurgy 1990,ed. E. R. Andreotti and P. J. McGeehan, MPIF, APMI, Princeton, NJ, 1990, v. 2.
5. K. Upadhya, Ceramic Bulletin, 1988, 1691-94.
6. K. Yamakazi, UC Davis, Private Communication, UCD, 1991.
7. S. A. Balankin, V. S. Sokolov and A. O. Troitzkiy, in PM'90, PM into the 1990's, International Conference on Powder Metallurgy [Proc. Conf.], London 1990, The Institute of Metals, London 1990, p. 37.
8. C.H. Liu and P.W. Kao, Scripta Metall. et Mater., 24 (1990) p. 2279.
9. A.S. Helle, K.E. Easterling and M.F. Ashby, Acta Metall., 1985, 2163.
10. I.I. Naumov, G.A. Ol'khovik and V.E. Panin, Izvestiya Akademii Nauk SSSR, Metally., No. 1 (1990) 152.
11. M. F. Ashby, Powders Compaction under Non-hydrostatic Stress States. An Initial Survey (1990).
12. D. L. Johnson, in Advanced Ceramics II, ed. S. Somiya, Elsevier, London, 1986, 1-6.
13. E. L. Kemer and D. L. Johnson, Am. Ceram. Soc. Bull., 64, (1985) 1132.

Section III

PLASMA HEAT TRANSFER - A KEY ISSUE IN

THERMAL PLASMA PROCESSING

E. Pfender

Department of Mechanical Engineering and
ERC for Plasma-Aided Manufacturing
Minneapolis, MN. 55455

Abstract

Plasma heat transfer plays a key role in most thermal plasma applications, as for example in plasma metallurgy, arc and plasma spraying, plasma waste destruction, plasma CVD, plasma synthesis, and plasma sintering. Similar as in the case of conventional heat transfer, the properties of the boundary layer separating a plasma from a wall or electrode determine the mechanisms as well as the magnitude of heat transfer. A plasma boundary layer, however , is much more complex than an ordinary boundary layer due to the existence of charged particles and the extremely steep gradients across such a boundary layer which may cause severe deviations from chemical as well as kinetic equilibrium. Plasma heat transfer will be discussed for (a) the case of a wall on floating potential (no net current flow), (b) the related case of heat transfer to particulates injected into the plasma, and (c) the case of heat transfer to an electrode (anode). Results of modeling efforts will be compared with available experimental data.

Plasma Synthesis and Processing of Materials
Edited by Kamleshwar Upadhya
The Minerals, Metals & Materials Society ©1993

Introduction

An effective utilization and exploitation of thermal plasmas requires a thorough understanding of the heat transfer process from a plasma to a solid or a liquid. This process is much more complicated than in the case of an ordinary gas. The existence of free electrons and positive ions in a plasma and the steep gradients of the plasma parameters, particularly in the vicinity of walls or electrodes, give rise to a number of effects which are still poorly understood. The interaction with relatively strong radiation fields which are typical for thermal plasmas, complicates the situation further.

Research in the field of plasma heat transfer received a strong impetus by plasma applications related to space flight and reentry simulation. Relevant literature references are summarized in ref. 1.

Over the past years there has been renewed research interest in thermal plasma heat transfer precipitated by applications in material processing. Materials processing in thermal plasma must be viewed in the context of much broader technology trends.

There is no question that materials and materials processing will be one of the most important technical issues as we approach the turn of the century. This will not be restricted to the development of new materials, but will also include the refining of materials, the conservation of materials (by hard facing, coating etc.), and the development of new processing routes which are more energy efficient, more productive, and less damaging to our environment. Thermal plasma processing will play an important role in these developments. Its potential for developing new materials-related technologies is increasingly recognized and many research laboratories all over the world are engaged in advancing the frontiers of our knowledge in this exciting field. An interesting example of the utility of plasma processing has been recently demonstrated in connection with a breakthrough in the field of diamond CVD. It has been shown that the highest deposition rates can be obtained in thermal plasmas [2,3].

In this paper, results of experimental and analytical plasma heat transfer studies, conducted in the High Temperature Laboratory, Heat Transfer Division, University of Minnesota will be summarized. No attempt will be made to give a comprehensive literature survey. Only those publications will be mentioned which are directly related to the material covered in this paper.

In the experimental work electric arcs and high frequency (r.f.) discharges have been used as a convenient means for generating plasmas in a temperature range from approximately 8×10^3 to 3×10^4K and electron densities ranging from 10^{16} to 10^{18} cm^{-3}. These plasmas can be described as continua since most of the studies refer to atmospheric pressure and the temperature is usually high enough so that the plasmas may be considered as thermal plasmas, i.e. they approach LTE. The cooling effect caused by walls and electrodes bordering the plasma or by particulate matter injected into the plasma extends, generally, into the plasma in a thin layer only which can, in the usual terminology be described as a boundary layer.

Analytical investigations in plasma heat transfer are almost entirely restricted to laminar flow because the degree and the nature of turbulence in the plasma flow is only poorly known.

The paper is divided into 4 sections. The first section reviews some basic aspects of plasma heat transfer, including a brief discussion of the various mechanisms which may contribute to the total heat flow. In the second section results of heat transfer studies to surfaces will be reviewed assuming that these surfaces are not subject to a net current flow. This situation prevails in all kinds of plasma generating devices in which the plasma is in "contact" with walls or other structural elements of the device (excluding electrodes).

The third section which is closely related to the second one refers to plasma application in which materials are injected into plasmas as, for example, in plasma spraying or plasma spheroidization. If there is a substantial flow of electric current to a surface, heat transfer may be dominated by the current flow. This situation will be discussed in the last section which deals with electrode heat transfer, in particular with heat transfer to the anode in high intensity arcs.

Basic Considerations

For a proper assessment of plasma heat transfer it is useful to evaluate qualitatively how much heat transfer in a plasma is expected to differ from heat transfer in an ordinary gas at low temperature.

The heat transfer phenomenon and, in particular, the thermal boundary layer are well understood for a cool, solid body immersed in a laminar hot gas stream in which no chemical reactions occur. In this simple situation, the heat transfer can be predicted by a dimensionless parameter, the Nusselt number

$$Nu = f(Pr, Re) \tag{1}$$

(Pr = Prandtl number, Re = Reynolds number). With some modifications, similar relationships hold for dissociating gases, provided that the Lewis number which describes the diffusion of the species is close to unity.

If one now considers heat transfer in a temperature range high enough for ionization to occur, one might expect a strong increase of the heat transfer coefficient because the free electrons contribute strongly to the thermal conductivity as they do in a metal. But this is not the case--at least not on a cold catalytic wall--because its surface is always separated by a cool less ionized layer from the hotter part of the boundary layer and from the main stream. This layer will be thicker when chemical equilibrium exists than for a chemically frozen boundary layer as shown in Fig. 1. Such a layer acts as a thermal insulator and prevents a large increase in heat transfer. From a purely theoretical point of view, only on a completely non-catalytic surface an electron density with finite values would exist throughout the boundary layer and considering electronic conduction only, a strong increase of the heat transfer to the surface would have to be expected.

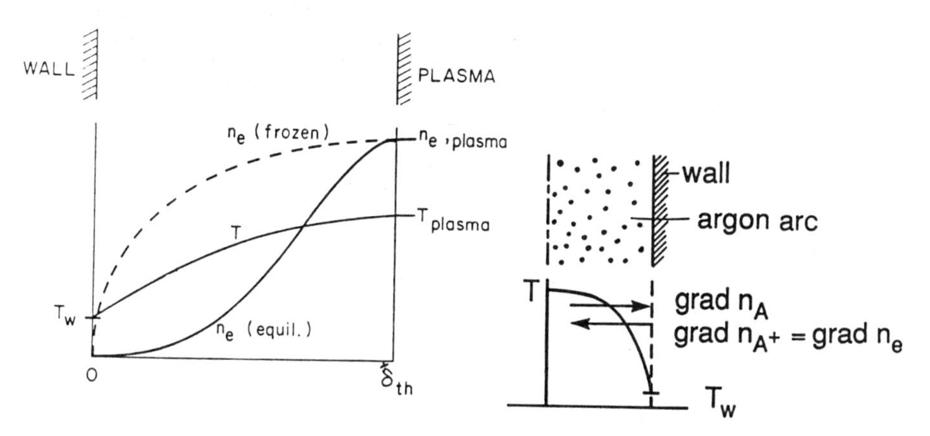

Fig. 1: Schematic of a plasma-wall boundary layer

Fig. 2: Temperature distribution in an argon arc (schematically) and corresponding density gradients

There is another factor involved in heat transfer from a plasma which tends to reduce the heat flux. In the interdiffusion of electrically charged and uncharged particles in an ionized gas, the electrons and ions move as if they were joined together (ambipolar diffusion). The diffusion coefficient D for this process is approximately a factor of two larger than the diffusion coefficient for the corresponding neutral particles. It therefore maintains its order of magnitude in the Lewis number

$$Le = \frac{\rho C_p D}{\kappa} \tag{2}$$

(ρ = density, C_p = specific heat at constant pressure, D = diffusion coefficient, κ = thermal conductivity). The thermal conductivity in a plasma, however, increases possibly by an order of magnitude or even more. This causes the Lewis number to assume values smaller than one. Therefore, a correction term has to be applied to the heat transfer coefficient which reduces the Nusselt number [1].

These considerations demonstrate that the Nusselt number and the heat flux to the surface are not as strongly influenced by the presence of electrons as one would expect and that the heat flux in a boundary layer at composition equilibrium will be smaller than in a frozen boundary layer. Experiments which verify these conclusions will be discussed in later paragraphs.

Another mechanism which is of minor importance in an ordinary gas at low temperature may play an important role if plasmas are considered. A thermal plasma is a strongly radiating gas. In general, the integrated radiation intensity is a function of temperature and pressure for a given plasma. For the temperature range which is of interest in this review, the radiation intensity increases strongly with temperature and also with pressure especially if metallic species are part of the plasma. For extremely high pressures (> 100 atm) the radiation intensity approaches that of blackbody.

Unfortunately, there is no general description of radiative heat transfer from a plasma to a surface because this transfer depends strongly on the nature of the plasma and on the entire configuration. Only in extremely simple situations of optically thin plasmas, generated from pure, monatomic gases, reasonably accurate values of the emission coefficient may be available for a prediction of radiative heat transfer.

In applications which utilize uncooled electrodes (for example carbon or graphite) radiative heat transfer from the electrodes to the material to be heated by the plasma may play an important role in the overall energy balance. Since steep gradients and chemical reactions (dissociation, ionization and the reverse reactions) are characteristic features of thermal plasmas, these features will give rise to heat and mass fluxes even if the plasma is macroscopically at rest. The most important contributions of mass fluxes across plasma-wall boundary layers are due to concentration and temperature gradients (ordinary and thermal diffusion, respectively). In term of heat transfer there are three contributions, i..e. heat conduction, enthalpy inter diffusion, and transport by the diffusion thermo effect. A more detailed description of these effects are given elsewhere [4].

It should be emphasized that for many cases of practical importance, thermal diffusion and the diffusion thermo effect may be neglected with respect to the other transport mechanisms. This, however, still leaves two terms in the heat flux equation which for the simple case of a two-component mixture (fully ionized plasma) may be written as

$$\vec{q} = -\kappa_{I=0}\,\mathrm{grad}T + (h_1 - h_2)\,\vec{I}_1 \tag{3}$$

where $\kappa_{I=0}$ is the heat conductivity for pure conduction (without mass fluxes), h_1 and h_2 are the enthalpies of component 1 and 2 respectively, and $\vec{I}_1$ is the mass flux of component 1. The role of the second term in Eq. (4) may be illustrated by considering the simple example of a high intensity argon arc enclosed in a tube which is kept at a relatively low temperature

T_W (Fig. 2). Because of the high degree of ionization in the hot core of the arc there will be density gradients of argon atoms and ions (electrons) as indicated in Fig. 2. These gradients will give rise to mass fluxes of atoms and ions, i.e.

$$\vec{I}_A = -L_1 \operatorname{grad} h_A \tag{4}$$

$$\vec{I}_{A^+} = -L_2 \operatorname{grad} h_{A^+} \tag{5}$$

where h_A and h_{A+} are the number densities of argon atoms and ions, respectively and L_1 and L_2 are coefficients. But there will be no net mass fluxes ($\vec{I}_A = -\vec{I}_{A^+}$). Since the opposite equal mass fluxes carry different enthalpies h_A and h_{A+}, there will be a net heat flux to the confining wall so that the total heat transfer may be written as

$$\vec{q} = -\kappa_{I=0} \operatorname{grad} T + (h_{A^+} - h_A)\,\vec{I}_{A^+} \tag{6}$$

and $$q^* = h_{A^+} - h_A \tag{7}$$

is known as the heat of transition which is in this case the excess enthalpy which the ions and electrons carry with respect to the neutral atoms. This enthalpy is essentially the ionization energy. In such an arc there is a continuous flow of ions and electrons by ambipolar diffusion towards the wall where electrons and ions recombine releasing their ionization energy and the neutral atoms diffuse simultaneously in the opposite direction so that there is no net mass flow. For a detailed analysis of the heat transfer situation in a plasma flowing towards or over the wall, the equations for mass and energy fluxes have to be introduced into the boundary layer equations [4].

Heat Transfer to a Wall on Floating Potential

In order to experimentally assess the effect of ionization on heat transfer to a body immersed into or in contact with a plasma, either small diameter wires have been swept through a plasma generated by a free-burning arc or a thin water-cooled tube has been moved step by step through the plasma. Both methods will be briefly discussed without experimental details which have been presented elsewhere [5, 6]. The underlying principle for obtaining local heat fluxes is the same for both the sweeping wire technique and the water-cooled probe.

Sweeping Wire Technique

The wire probe consisting of W or Pt is swept with constant velocity through the plasma generated by a free-burning arc in argon atmosphere as shown schematically in Fig. 3. From the complete history of the resistance change of the wire, local heat fluxes to the wire averaged over its circumference and the temperatures of the wire can be derived. The rate of change of the resistance is related to the heat flow to the wire whereas the magnitude of the resistance reflects the temperature of the wire. With the known properties of the wire material, the heat flow to the wire Q(x) is obtained where x denotes the distance of the probe center from the plasma center. The measured distribution Q(x) may be converted into local heat fluxes q(r) using the Abel inversion technique similar to that utilized in spectroscopy. A prerequisite for the application of this method is rotational symmetry of the plasma. Since $T_{plasma} \gg T_{wire}$ the plasma temperature determines the heat transfer independent of the wire temperature. The heat transfer from the plasma to the probe is adequately recorded if external and not internal conditions govern the heat transfer process. By specifying an external and internal resistance for heat transfer, one can show that the Biot number $Bi = R_{int}/R_{ext} < 0.03$ for the conditions in this experiment.

The effect of radiative heat transfer from the arc to the probe has been estimated. Taking ff- and fb-radiation into account as well as the shape factor of the wire with respect to the

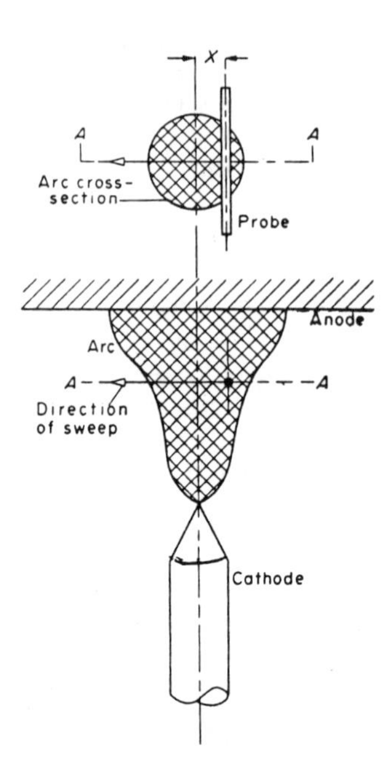

Fig. 3: Schematic of the sweeping wire technique applied to a free-burning arc

luminous arc, radiation contributes between 4-10% of the total heat transfer to the wire. For the evaluation of the data, the free stream properties of the arc plasma surrounding the probe are required. Plasma temperatures and velocities have been measured for 200 and 300 A arcs in argon at p = 1 atm and electrode gaps of 10 and 11.2 mm, respectively [7, 8].

Close to the cathode the plasma temperature exceeds 2 x 10^4K and the corresponding degree of ionization is almost 100%. Velocities induced by the cathode jet exceed 600 m/s at this location. At a distance 5.5 mm from the cathode, the axis temperature in the arc drops to approximately 1.5 x 10^4K and the associated degree of ionization is close to 50%. The heat transfer measurements which will be discussed in the following have been taken over this range i.e. from approximately 50-100% ionization.

In the case of the sweeping wire technique, bare metal wires (W and Pt) are swept with constant speed through the arc plasma in planes perpendicular to the arc axis at distances of approximately 1.5, 3, 4 and 5.5 mm from the cathode tip. In this way, a wide range of plasma temperatures and velocities can be covered by the probe sweeps. Considering convective heat transfer only, the Nusselt number is related to the Reynolds number by $Nu_d \sim Re_d^n$. For Reynolds numbers in the range from 10 to 50 which are typical for these experiments, the exponent n is approximately 0.5 so that $Nu_d \sim d^{1/2}$ which agrees with the trend of the data. The effect of the sweeping rate has been eliminated by extrapolating the data to zero sweeping rate. A set of typical data is shown in Fig. 4 for distances from 1.55 to 5.4 mm from the cathode. The spread of the curves for various sweeping velocities indicates that the assumption of an equilibrium boundary layer is not consistent with this behavior. This finding is corroborated by the following consideration of characteristic time constants. The characteristic time for energy transport to the sweeping probe according to its velocity is $T_{probe}= 10^{-3} - 4x10^{-3}$ seconds. The corresponding time constants for an equilibrium and frozen boundary layer are listed below.

State of boundary layer	Characteristic time constant
Equilibrium	$T_{equil} = \Lambda^2{}_{equil}/\alpha = 10^{-7} - 10^{-5}$ sec
Frozen	$T_{frozen} = (10r_p)^2/D_{amb} = 10^{-5} - 10^{-3}$ sec

Λ represents a characteristic diffusion length, α the thermal diffusivity, r_p the probe radius and D_{amb} the ambipolar diffusion coefficient. Since T_{frozen} may reach the same order of magnitude as T_{probe}, the observed dependence of the heat flux to the probe on the sweeping rate is evidence that the boundary layer is not in an equilibrium state. According to this finding, the curves are extrapolated to zero sweeping rate as indicated by the dashed lines in Fig. 4.

Water-Cooled P robe

As previously mentioned, the principle for obtaining local heat fluxes (Abel inversion) is the same as for the sweeping wire technique. The water-cooled probe, however, utilizes a calorimetric procedure for acquiring the raw data which yields higher accuracies of these data compared to the sweeping wire technique. In addition, biasing of the tube with various potentials is a straightforward procedure which allows the determination of the effect of electric currents on heat transfer. Local values of heat fluxes and current densities can be measured simultaneously which facilitates comparison with analytical correlations. In order to determine whether or not there is any effect associated with the catalyticity of the wall, the tube has been coated with a thin layer (20 μm) of pyrex glass. Figure 5 shows a schematic of the probe in the mid-section of the arc plasma.

The heat transfer, Q(x), to the bare and pyrex coated water-cooled copper tube has been determined for arc currents of 200 and 300 A and the measured distribution, Q(x), are then inverted by the Abel inversion technique. In this way, circumferentially averaged local heat fluxes, q(r), are obtained. Since any probe assumes an average floating potential in the plasma, the inversion of the heat flow does not result in the heat flux to a locally floating probe, but represents some effective heat flux, $q_o(r)$. Determination of heat fluxes to a truly locally floating surface requires a series of heat flow measurements with various biasing potentials [8].

Local heat flux and effective heat flux for the pyrex and metal tube surfaces are shown in Fig. 6 for a plane 4.6 mm below the anode surface. For comparison, results obtained from the sweeping wire technique for the 200A arc using bare tungsten wires are also included in this figure as dots. The agreement between the results obtained with these two different methods is within a few percent in the probe center but deviates appreciably for $r > 0.5$ mm. Since the heat flux to an insulating surface (pyrex) is within experimental accuracy identical to that of a floating copper surface for the free-stream conditions of this experiment, it may be concluded that under these conditions all walls whether or not conducting behave as catalytic walls for recombination of ions with electrons. This conclusion is further corroborated by analytical considerations presented in Ref. 8.

Based on thermal conductivities and diffusion coefficients presented by Fay and Kemp [9] the temperature distribution and the electron density in the region in front of the probe which is governed by conduction and diffusion can be calculated for the two extreme cases of chemical equilibrium and frozen chemistry [8].

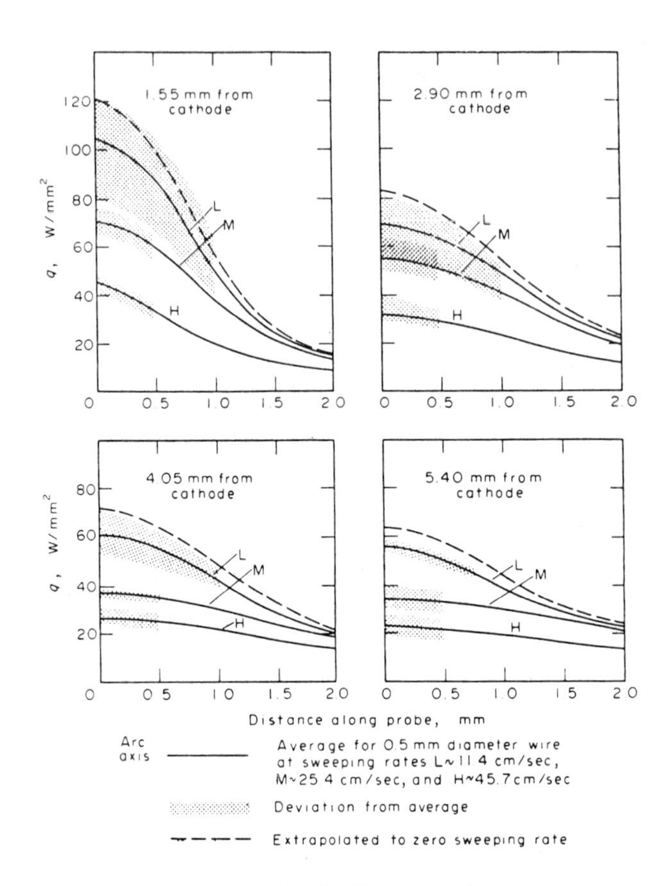

Fig. 4: Effect of probe diameter and sweeping rate on heat transfer to bare tungsten wires

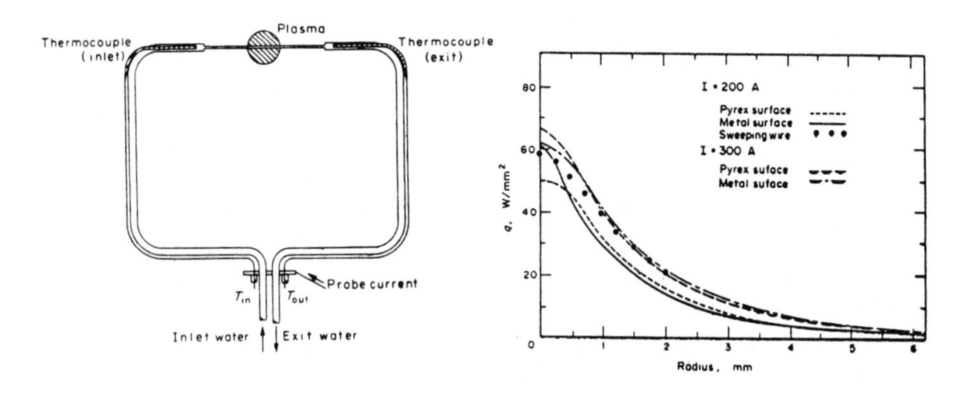

Fig. 5: Schematic of the water-cooled probe

Fig. 6: Local and effective heat fluxes along various probes

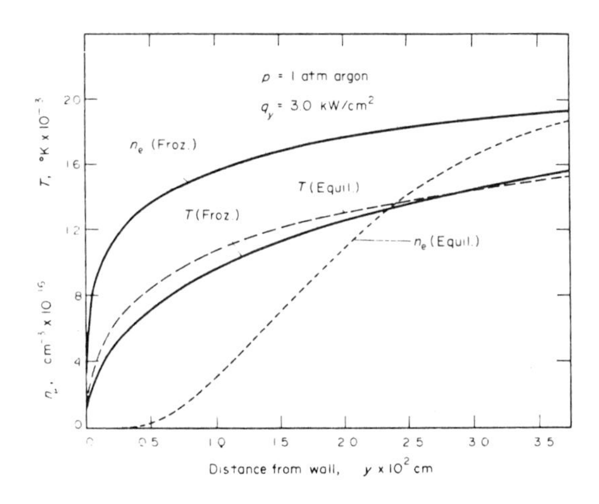

Fig. 7: Calculated electron density and temperature distribution in a plasma-wall boundary layer

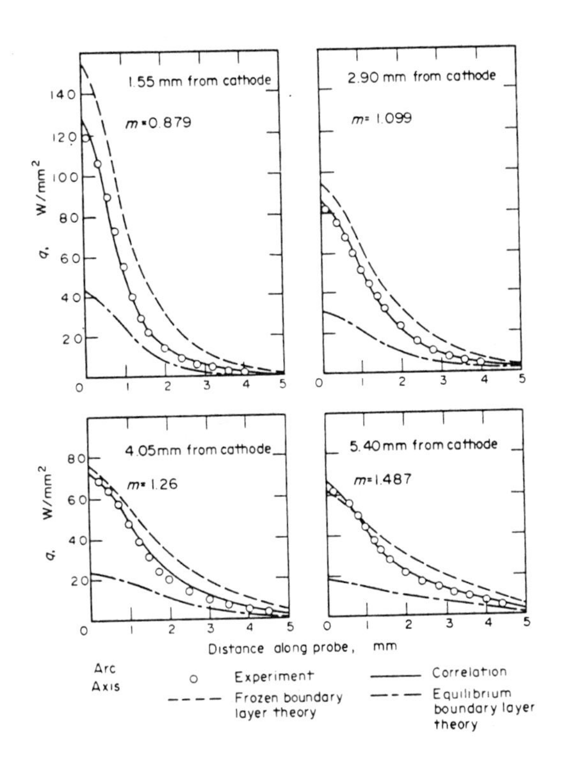

Fig. 8: Comparison of sweeping wire data with chemical equilibrium and frozen chemistry limits

Results for a wall heat flux of 3 kW/cm^2 are presented in Fig. 7. As expected, the electron density distribution changes significantly for frozen conditions in the boundary layer which has a thickness in the order of 10^{-2} cm. A conservative estimate of the diffusion transit time through the boundary layer and of the recombination time demonstrates that they are of the same order of magnitude which clearly indicates that deviations from chemical equilibrium have to be expected. Further evidence for frozen chemistry is demonstrated by the results of the sweeping wire technique. An upper and a lower limit for the heat transfer to the wire is obtained by assuming either a completely frozen or an equilibrium state. Figure 8 shows the corresponding comparison which, again, indicates that frozen conditions are closely approached.

In the previous discussion, possible deviations from kinetic equilibrium in the plasma-wall boundary layer have not been considered, i.e. it has been assumed that kinetic equilibrium prevails in the boundary layer ($T_e=T_h$). This, however, is not the case as shown by modeling and electric probe measurements close to the wall of a wall-stabilized arc [10, 11]. Both modeling and measurements show that the electron temperature in the wall boundary layer stays above 9,000 K (Fig. 9, 10) up to the wall and it seems that the electron temperature gradient at the wall approaches zero. Therefore heat conduction due to the electron gas will not be important, but recombination of ions and electrons at the wall may still be of importance as previously discussed.

Plasma-Particle Heat Transfer

Heat, mass, and momentum transfer which particulates experience in a thermal plasma flow are important for materials processing, including extractive metallurgy, plasma deposition, plasma synthesis and densification, and for plasma decomposition of toxic waste materials. In this paper, only the heat and mass transfer aspect will be considered.

In general, heat and mass transfer to and from a particle immersed into a plasma may be affected by

- unsteady conditions,
- modified transfer coefficients due to strongly varying plasma properties,
- vaporization and evaporation,
- non-continuum effects,
- radiation,
- particle shape,
- particle charging,
- combination of above effects.

The relative importance of these effects has already been discussed in the literature and is summarized in refs. 12 and 13. Only a few of them will be discussed in the following.

Thermal plasmas with temperatures in the order of 10^4K provide extremely high heating rates for injected materials as well as high quench rates for particles leaving the hot plasma. The latter aspect is of particular importance for plasma synthesis and for rapid solidification processes. Because of the high temperatures which are characteristic for thermal plasma reactors, the required processing time become very short, and this translates into relatively small reactors with high throughput rates.

In general, thermal plasma reactors are highly heterogeneous systems with large temperature and velocity variations. Over a distance of 10 mm, for example, the temperature may drop from 15,000 K to almost room temperature, or the velocity may drop from 500 m/s to almost zero. In such systems the injected particles must pass through the hottest region in the plasma with maximum possible exposure to complete the desired physical or chemical processes. In some applications, fast acceleration of particles in the plasma is desirable so that they accumulate high velocities before impinging on a target.

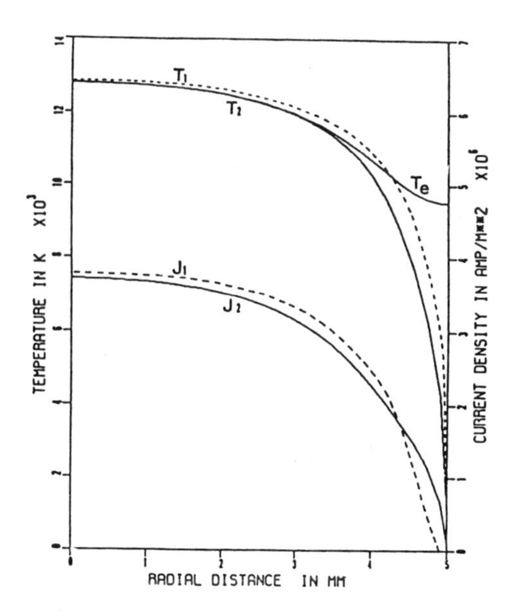

Fig. 9: Temperature and current density distributions in a fully developed argon arc at 1 atm and 200 A; 1 refers to 1-T model, and 2 refers to 2-T model

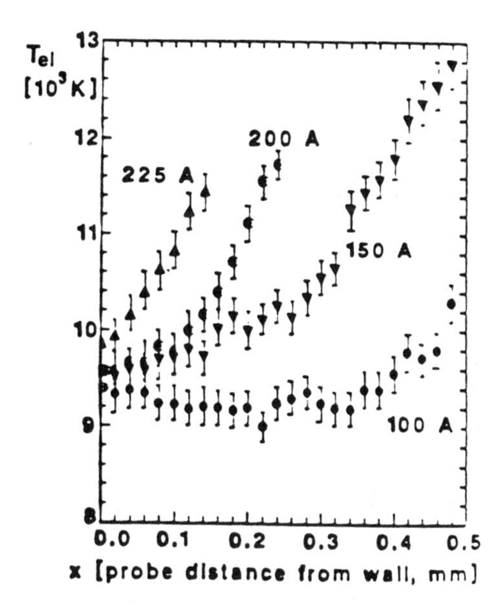

Fig. 10: Electron temperature proflies for various arc currents

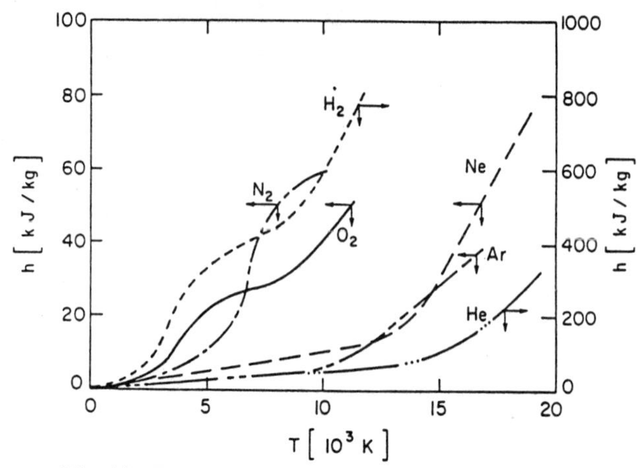

Fig. 11: Enthalpy of some monatomic and diatomic plasma gases

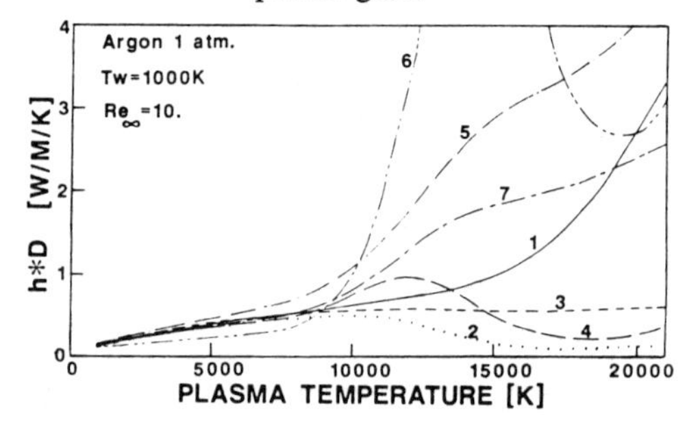

Fig. 12: Seven heat transfer coefficients vs. plasma temperature for plasma Reynolds number=10. (1) Lewis and Gauvin; (2) Fiszdon; (3) Sayegh and Gauvin; (4) Lee, Hsu, and Pfender; (5) Vardelle, Vardelle, Fauchais and Boulos; (6) Kalganova and Klubnikin; (7) Chen and Lin.

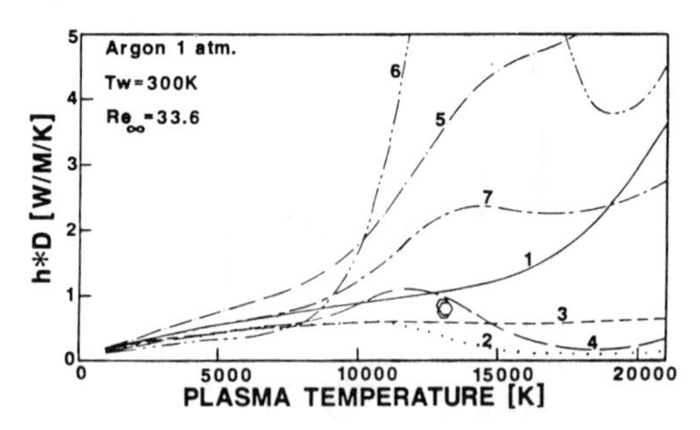

Fig. 13: Comparison against Petrie et al. experimental data for flow across cylinder. (1) Lewis and Gauvin; (2) Fiszdon; (3) Sayegh and Gauvin; (4) Lee, Hsu, and Pfender; (5) Vardelle, Vardelle, Fauchais, and Boulos; (6) Kalganova and Klubnikin; (7) Chen and Lin.

One of the key factors for satisfying these requirements is the control of the particle trajectories and of the heat and associated mass transfer between particles and plasmas. In spite of increasing efforts over the past 20 years, understanding of the interaction of particulate matter with thermal plasmas still remains incomplete.

For an understanding of the behavior of particulates injected into a thermal plasma, it is necessary to know the conditions to which the particulates will be exposed. The plasma environment is governed by the plasma composition, temperature, enthalpy, and velocity fields. The plasma enthalpy is strongly affected by the plasma composition as shown in Fig. 11. The presence of molecular species in the plasma gives rise to substantially higher enthalpies at a given temperature which, in turn, will affect heat transfer. In addition, the plasma composition may have strong effects on the transport properties. "Contamination" of plasmas with small percentages of low ionization potential materials (for example metals) may have a drastic effect on some of the transport properties especially on the electrical conductivity [14, 15]. Such "contaminations" with low ionization potential species is almost inevitable in plasma processing of particulate matter.

Heat transfer from plasmas to spherical particles has been extensively discussed in the literature (for references see ref. 12). There are, however, still large discrepancies in terms of heat transfer coefficients. For argon plasmas there is a general consensus up to 9,000 K beyond which wide deviations occur. For nitrogen plasmas, there is good agreement among various investigators up to 4,000 K with large discrepancies beyond this temperature. As an example, Fig. 12 shows heat transfer coefficients for the case of argon, and Fig. 13 comparisons with the few experimental data available [16]. The discrepancies among various authors are even more pronounced in the case of nitrogen plasmas shown in Fig. 14. It should be pointed out that these examples were only concerned with plasma heat transfer situations for a single monatomic gas (argon) and a single diatomic gas (nitrogen). However, in many practical plasma systems it is customary to use mixtures of two or more gases, especially in the case of plasma synthesis. There are indications that plasma heat transfer for gaseous mixtures is vastly different from pure gases. Chludzinski et. al. [17] have shown in an experimental study that the addition of just a small amount of nitrogen (8%) to an argon discharge will increase the heat transfer coefficient by 80%. This seems reasonable because frozen heat transfer theory predicts a strong increase of heat transfer due to diffusion of dissociation energy to the sphere's surface which increases the heat transfer coefficient.

At a first glance it seems desirable to add polyatomic gases to plasma systems, but as Capitelli et. al. [18] have shown for a constant power level, the plasma volume decreases when, for example, nitrogen is added to the discharge. This may result in reduced heating of injected powders because of the decreased residence time of powder particles in the smaller plasma volume.

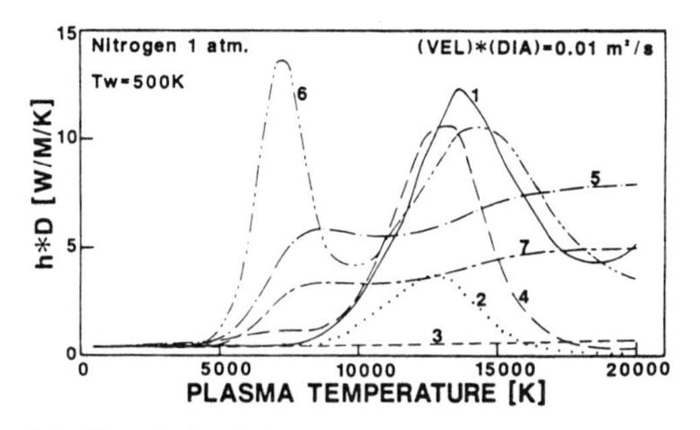

Fig. 14: Correlation behavior for nitrogen. (1) Lewis and Gauvin; (2) Fiszdon; (3) Sayegh and Gauvin; (4) Lee, Hsu, and Pfender; (5) Vardelle, Vardelle, Fauchais, and Boulos; (6) Kalganova and Klubnikin; (7) Chen and Lin.

Another effect which should be pointed out is associated with vaporization and evaporation of particulates. The experimental data included in Fig. 13 refer to particle surface temperatures less than or equal to the melting temperature of the solid, but substantially below the boiling temperature. This situation applies to applications such as plasma spraying, where the goal is to melt the particles. In contrast, for successful plasma chemical synthesis, complete evaporation of the injected powder is a primary requirement. As shown in ref. 19 the evaporation stage is by far the longest in thermal plasma processing. Heating of a particle from its initial temperature to melting, to fully melt the particle, and to then heat it beyond to nearly the boiling point is relatively rapid. Evaporation, however, requires an order of magnitude longer to complete.

Vaporization and evaporation are physical processes concerned with mass transfer across a liquid-vapor interface. Vaporization is defined as a mass transfer process driven by vapor concentration gradients existing between the free stream and the particle surface. In contrast, evaporation accounts for large amounts of mass transfer as the surface temperature reaches the boiling point.

At the interface between liquid and vapor phases, a heat balance is maintained and at the same time a bulk flow of material crosses the interface. For high mass transfer rates, the transfer coefficients become functions of the mass transfer rate, thus causing nonlinearities in the transport equations. For example, in ref. 19 it has been shown that the heat flux through a liquid-vapor interface is reduced due to the absorption of heat by the vapor. This, however, does not invalidate the definitions of the transfer coefficients.

Numerical simulation of a particle residing in a high-temperature surrounding has been performed based on the model described in ref. 20. The noncontinuum effect which will be discussed later on has been included. Two limiting cases for vaporization can be established. For Case I, mass vaporization (or mass diffusion) is taken into account [21] before the particle surface reaches the boiling point. In Case II the vaporization rate driven by the vapor pressure is determined by eqs. proposed in refs. 22 and 23.

The results of corresponding calculations are presented in Figs. 15 and 16. In Fig. 15 normalized temperatures are plotted as a function of time after the particle has been exposed to a thermal plasma. Particular attention is focused on the region where vaporization or evaporation occurs. Figure 15 indicates that the surface temperature reaches a "plateau" (Case II) at some temperature between the boiling point and continues to rise very slowly. For Case 1, there is no vaporization before the surface temperature reaches the boiling point. The results of Fig. 16 show that the total mass transfer through the interface is almost the same for both cases, which is an important finding. Additional studies have shown that the differences between the two limiting cases becomes less significant for higher free stream temperatures and/or lower boiling point materials.

Since the particle sizes used for plasma processing may be of the same order of magnitude as the molecular mean free path lengths in the plasma. This "rarefaction effect" may exert a strong influence on heat transfer.

The noncontinuum effect on heat transfer has been studied in ref. 24 in the temperature jump regime, resulting in a proposed correction:

$$\frac{q_{noncont}}{q_{cont}} = \frac{1}{1 + (Z^* / r_p)} \tag{8}$$

where Z^* is the jump distance and r_p the particle radius.

The noncontinuum effect becomes substantial for small particles. Therefore, it is crucial for modeling associated with thermal plasma processing [25] when small particles (< 20 μm) are

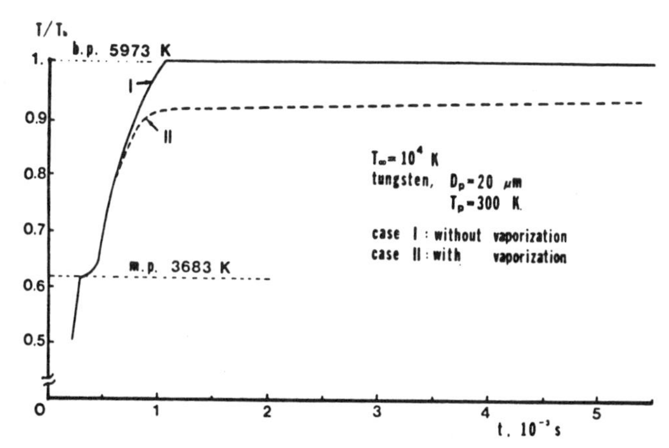

Fig. 15: Normalized temperature history of vaporization and evaporation. (Note: in Case I evaporation starts as the particle surface reaches the boiling point.)

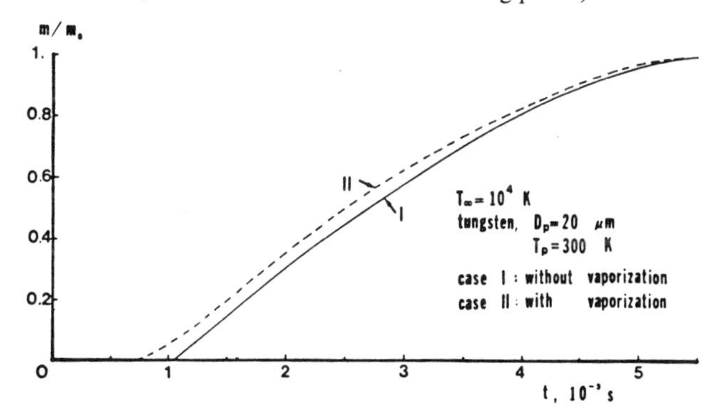

Fig. 16: Normalized total mass transfer from a particle (m_0=initial mass of the particle).

involved. The approach used in ref. 25 is based on the so-called temperature jump which is valid for Knudsen numbers in the range 0.001< kn <1.

As pointed out earlier, the boundary layer surrounding a particle is characterized by strong deviations from chemical as well as kinetic equilibrium. And, as just discussed, the noncontinuum effect will play an important role if small particles (< 50 μm) are considered. In an attempt to include all these effects in a unified theory for describing particle heat transfer [26] the contributions due to heat conduction of the heavy species has been separated from the energy transport by ambipolar diffusion of charged particles to the surface. The Knudsen effect has been accounted for by using Sherman's interpolation formula [27] for the conduction problem and its equivalent, based on electrostatic probe theory [28], for the charged particle transport to the surface. Fig.17 shows a typical set of results for $\tau_p = T_p / T_\infty = 0.1$, where T_p is the particle surface temperature and T_∞ the undisturbed plasma temperature. Figs. 17a to 17c indicate that the dominating heat transfer mechanism changes over a relatively small temperature interval (approximately 3,000K wide) from ordinary conduction of the heavy species to recombination. Included in the recombination part is the energy which the ions acquire in the sheath. Lowering of the pressure shifts the crossover to lower temperatures due to the higher degree of ionization at reduced pressures for a given temperature.

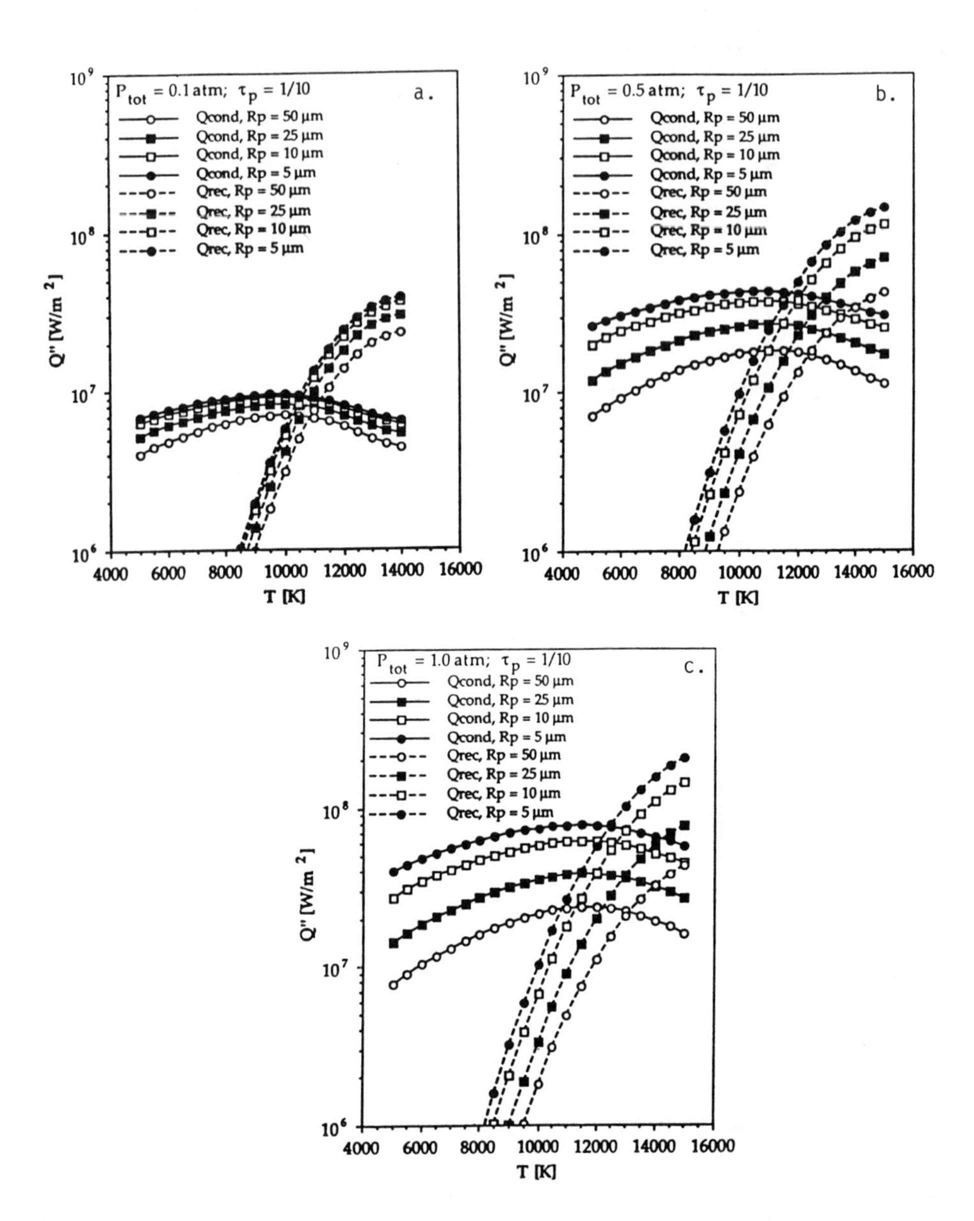

Fig. 17: Heavy species conduction (Q_{cond}) and ion current energy transport (Q_{rec}) per unit area to particles of Radius R_p vs undisturbed plasma temperature for a total pressure of (a) 0.1 atm, (b) 0.5 atm, (c) 1 atm.

Substantially more work is needed to clarify the effects of particle shape and particle charging on heat and mass transfer. Some discussion about these effects is included in ref. 29.

Anode Heat Transfer

The high pressure, high current (high intensity) arc represents the primary tool for producing thermal plasmas which are of increasing interest for chemical and material processing. In contrast to low intensity or low current arcs, high intensity arcs ($I > 50$ A) at atmospheric or higher pressures are characterized by strong macroscopic flows induced by the arc itself. Any variation of the current-carrying cross section of the arc leads to induced flows [30]. At sufficiently high currents and axial current density variations, flow velocities in the order of 100 m/s are observed. The cathode jet phenomenon in high intensity arcs is a typical example. In this case, the electron emission process at the cathode causes a contraction of the arc in front of the cathode, which, in turn, gives rise to a continuous magnetohydrodynamic (MHD) pumping action producing the well-known cathode jet. Similar effects may occur in the anode region of high intensity arcs, which is strongly affected by such flows induced by the arc itself.

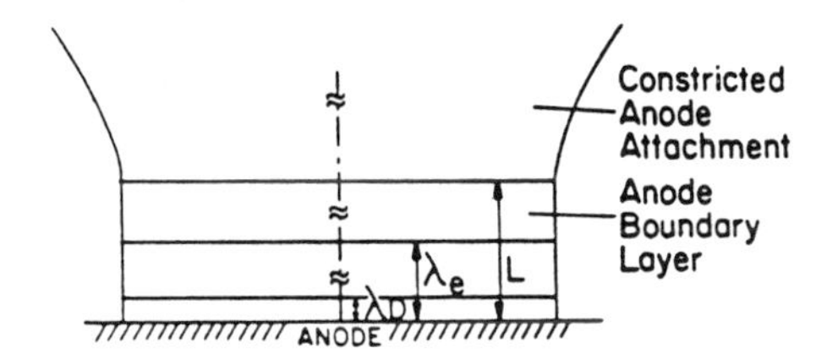

Fig. 18: Scale lengths of plasma in the anode boundary layer. L-anode boundary layer thickness; λ_e - electron mean-free path; λ_D - Debye length.

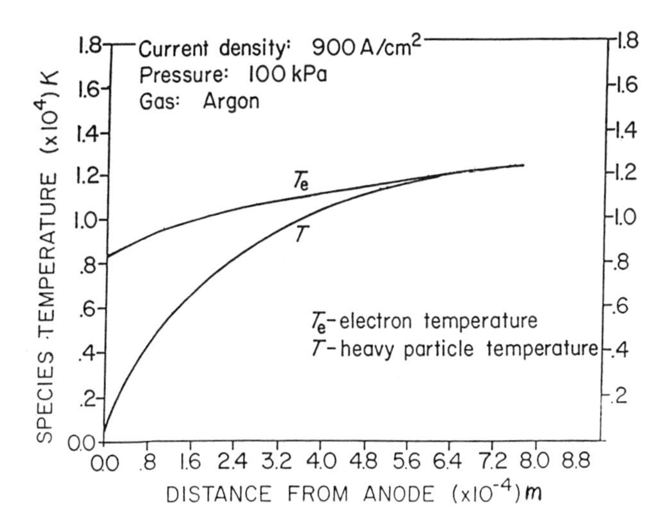

Fig. 19: Electron and heavy particle temperature in the anode boundary layer of a high intensity arc

The anode region may be divided into several zones. A flow-effected zone, as mentioned before, followed towards the anode by a layer in which the presence of the relatively cold anode is felt and which is characterized by steep gradients of temperature and particle densities. In the usual terminology, this layer may be designated as a boundary layer. At the bottom of this boundary layer is the sheath overlying the anode surface as sketched in Fig. 18.

A one-dimensional analysis of this anode boundary layer reveals substantial deviations from LTE [31]. The temperature of the heavy species approaches the temperature of the anode in the immediate vicinity of the anode surface, whereas the electron temperature remains sufficiently high to ensure the required electrical conductivity as shown for a typical example in Fig. 19. Temperature and density gradients in the anode boundary layer contribute substantially to the electric current flow so that the potential drop across the boundary layer may become negative. Eq. (10) describes the current flow of this situation

$$j=\sigma_e\left(E+\frac{1}{en_e}\frac{dp_e}{dx}\right)+\phi\frac{dT_e}{dx} \tag{9}$$

where j is the current density, T_e the electron temperature, σ_e the electrical conductivity, E the electric field strength, n_e the electron density, p_e the partial pressure of the electron gas, ϕ the thermal diffusion coefficient, and x the distance from the anode surface for the one-dimensional model.

Electric probe measurements at and close to a plane water-cooled anode surface in atmospheric pressure, high intensity argon arcs for different arc configurations which, in turn, result in two distinctly different anode arc roots (diffuse and constricted), arc in qualitative agreement with the previous analysis [32]. In the case of a constricted anode arc root, the potential drop across the anode surface is positive, whereas in the case of a diffuse anode arc root, a negative anode fall has been found [32]. This fact has a strong bearing on the heat flux carried by the electric current. In the case of a positive anode fall, the heat flux may be described by the conventional model

$$Q_{eL}=j\left[\frac{5}{2}\frac{kT_e}{e}+W+U_a\right] \tag{10}$$

and in the case of a negative anode fall one finds

$$Q_{eL}=j\left[\left(\frac{5}{2}+\frac{e\phi}{k\sigma_e}\right)\frac{kT_e}{e}+W\right] \tag{11}$$

where W is the work function of the anode material, U_a the anode fall, and k the Boltzmann constant.

Because of the general importance of these results, a more detailed study of the anode boundary has been undertaken, considering a free-burning high intensity arc in argon atmosphere [33] with emphasis on the heat transfer characteristics. A schematic of the experimental setup is shown in Fig. 20. Measurements of electron temperatures and electron and ion densities have been performed with the probe flush with the anode surface. A typical probe characteristic is shown in Fig. 21. The electron temperature follows directly from the slope of this characteristic and a set of electron temperatures as a function of the arc current for various arc gaps are shown in Fig. 22. It should be pointed out that both, increase of the current and decrease of the electrode gap enhance the velocity of the cathode jet in front of the anode, i.e. they reduce the thickness of the anode boundary layer resulting, as expected, in a steeper electron temperature gradient at the anode. The electron temperatures throughout the boundary layer up to the anode surface stay above 10^4K. The same probe can be used to

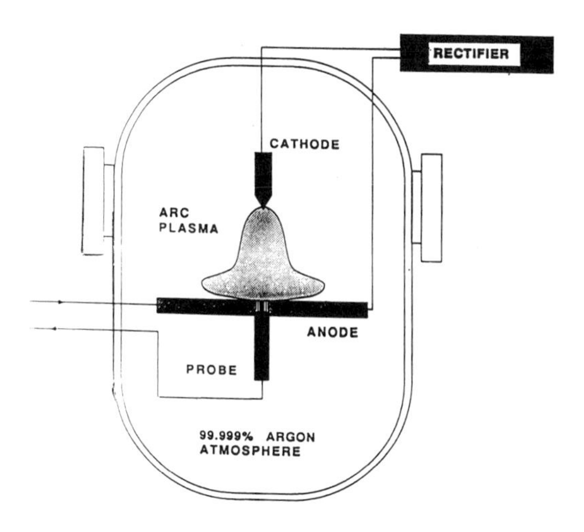

Fig. 20: Schematic of the experimental facility

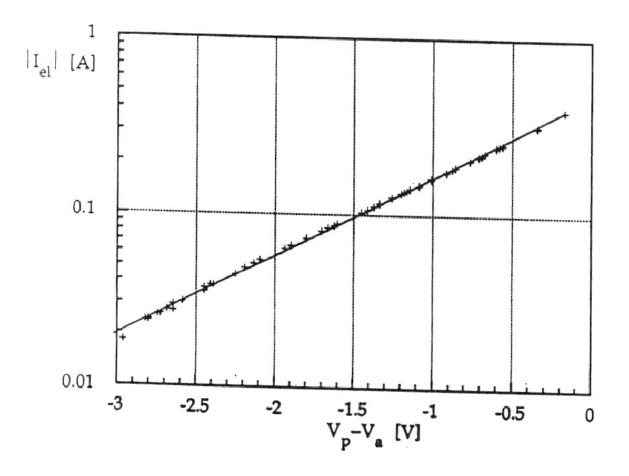

Fig. 21: Probe characteristic

determine current densities in the arc axis at the anode (maximum current densities). Typical results are shown in Fig. 23 which indicates a clear non-monatomic dependence on the overall arc current I_{arc}, displaying a minimum which, for the shorter gaps, is very pronounced. The increase in J_e, for I_{arc} falling below the value at which the minimum occurs, can be interpreted in terms of a constriction of the current carrying cross-section associated with the need for the arc to maintain sufficiently high temperatures and electrical conductivity in the core. Larger J_e values compensate by ohmic heating the losses by conduction due to the steeper radial, and possibly axial, gradients induced by the constriction typical of low current regimes. This current constriction at low arc current appears more severe at shorter gaps.

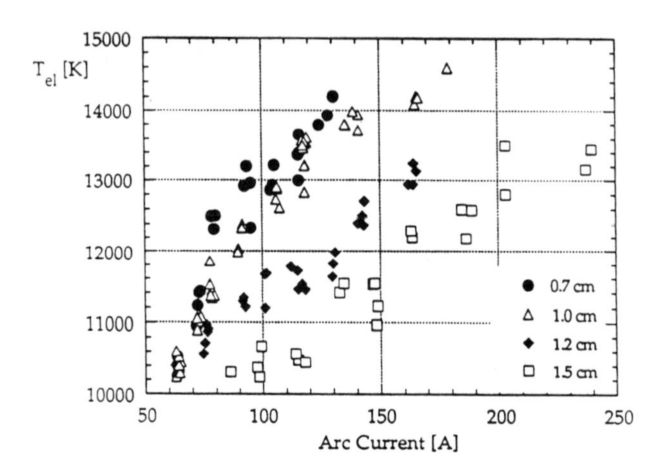

Fig. 22: Electron temperatures at the anode

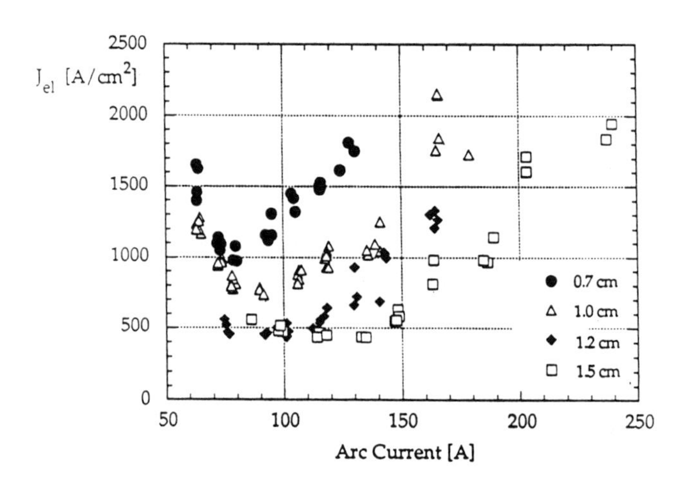

Fig. 23: Electron current densities at tha anode

With increasing I_{arc}, J_e eventually increases due to the growing influence of the cathode flow: this produces higher temperatures in front of the anode, associated with larger equilibrium densities, and in a reduction of the characteristic length over which ionizational nonequilibrium takes place. These two effects compound to produce an effective boundary layer structure at the anode, where large gradients exist over a distance much shorter than the characteristic length of the arc and where the resulting diffusion forces play a fundamental role as the current driving mechanism. As the arc current is increased, the boundary layer becomes thinner, and the resistance to the current flow decreases. The position of the current density minimum shifts to higher I_{arc} for larger gaps, indicating again that the cathode induced flow governs the establishment of a diffusion dominated anode attachment.

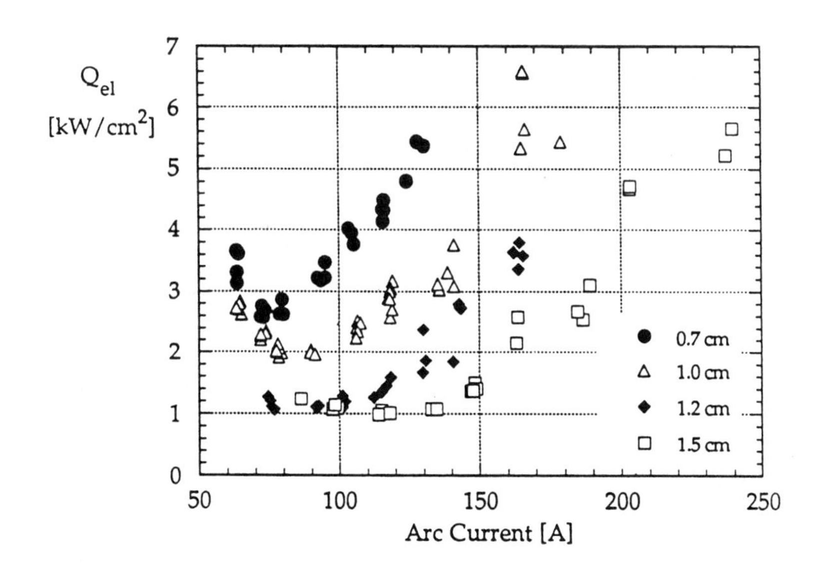

Fig. 24: Heat flux at the anode

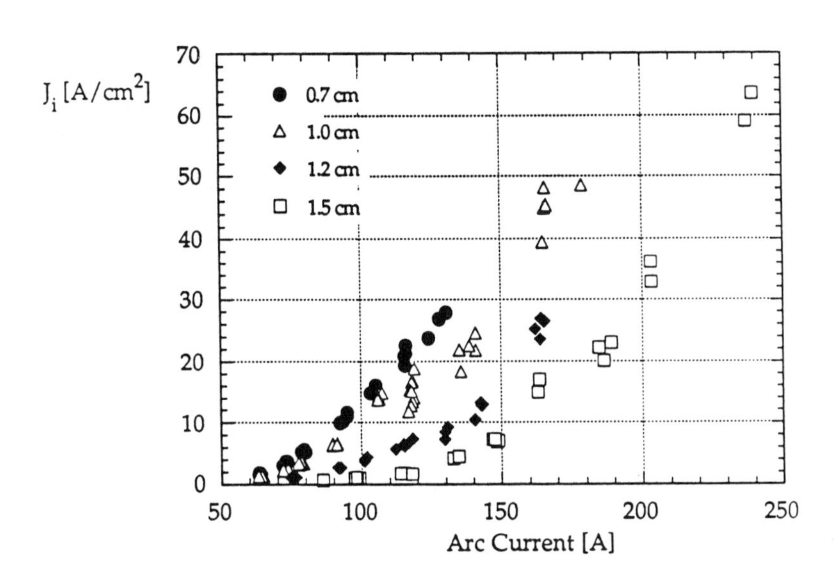

Fig. 25: Ion current density at the anode

The heat flux carried by the electrons according to Eq. (11) is shown in Fig. 24 indicating a similar behavior as the electron current density. This finding is of practical importance for the design of arc plasma devices. By keeping the anode boundary layer thin, the specific heat flux at the anode arc root may be reduced.

As previously mentioned, the gradients in the case of a very thin boundary layer become extremely steep so that electric currents driven by these gradients reach the anode. In the case of electron currents, this may result in negative anode falls as previously discussed. But not only electron currents, also ion currents are driven towards the anode establishing an electron retarding positive sheath in front of the anode. Since this sheath is more positive than the anode, it explains the negative anode fall and the associated reduction in the electron current reaching the anode. Some of the gradient driven positive ions will actually reach the anode resulting in an ion current according to Fig. 25. Although this ion current is approximately two orders of magnitude less than the electron current, it plays an important role for the interpretation of the interaction between boundary layer and anode.

Besides heat transfer to the anode by the current flow, heat transfer by convection and radiation has to be considered. Based on studies in a free-burning argon arc for currents between 50 and 350 A, Sanders [32] concluded that approximately 50% of the anode heat flux is due to the current flow, approximately 45% is due to conduction/convection, and the remainder (5%) represents radiative heat transfer. This distribution, however, depends strongly on the arc configuration, the arc gas composition, the power input, and on the pressure. In configurations in which the cathode jet impinges on the anode, convective heat transfer is enhanced and it may even dominate anode heat transfer at higher current (power) levels [34]. Radiative heat transfer which is relatively small in the case of atmospheric

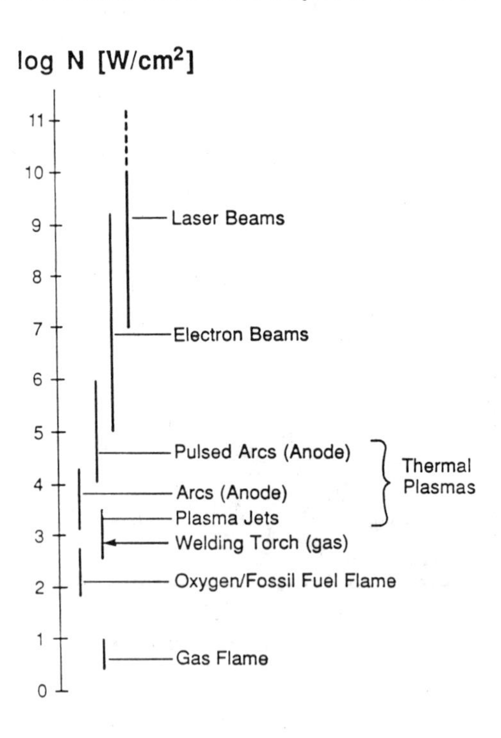

Fig. 26: Specific heat fluxes

pressure argon arcs may contribute substantially to anode heat transfer if the plasma contains metallic or other vapor components with low lying energy levels (arc lamps). And at high pressures ($p > 50$ atm), radiation may become the dominating heat transfer mechanism.

In general, total anode heat fluxes in steady state arcs may be in the range from 1 to 20 kW/cm^2 and they may reach values in the MW/cm^2 range in pulsed arcs [35] as shown in the schematic diagram of Fig. 26 in which anode heat fluxes are compared with heat fluxes produced in flames, plasma jets, electron and laser beams.

Summary and Conclusions

After some basic considerations, the first part of this paper has been devoted to plasma heat transfer studies for situations in which a wall on floating potential borders a plasma. Two experimental techniques employed for these studies clearly demonstrate that the boundary layer separating the wall from the plasma approaches a chemically "frozen" state and the experimental findings are in reasonable agreement with analytical predictions. Electric probe studies in such boundary layers show strong deviations from kinetic equilibrium ($T_e >> T_h$).

For the related situation of heat transfer to particulates injected into the plasma, there are a large number of possible effects which may influence heat transfer. This may be one of the reasons why there is still a large discrepancy among plasma heat transfer coefficients reported by different authors. It is anticipated that present worldwide research efforts in this area will resolve this problem in the near future.

Finally, studies of the anode boundary layer in a high intensity arc reveal strong deviations from kinetic equilibrium ($T_e >> T_h$) with reasonable agreement between analytical predictions and experimental results. A thin anode boundary layer and correspondingly strong gradients across this boundary layer drive electric currents across the boundary layer resulting in the formation of an electron retarding sheath in front of the anode and in negative anode falls. Anode heat transfer at the location of the anode arc root is reduced in this situation whereas a relatively thick, cold boundary layer leads to a constricted arc root and higher specific heat fluxes.

Acknowledgments

Prof. Xi Chen, Tsinghua University of Beijing, PRC and many former graduate students (s. references) contributed to the material presented in this paper. NSF as well as DOE supported this work over many years.

References

1. E.R.G. Eckert, and E. Pfender, Advances in Plasma Heat Transfer, in Advances in Heat Transfer, (New York, NY: Academic Press, 1967) 4.

2. N. Ohtake and M. Yoshikawa, "Diamond Film Preparation by the Discharge Plasma Jet Chemical Vapor Deposition in the Methane Atmosphere ", J. Electrochem. Soc., 137 (1990) 717.

3. E. Pfender, Q.Y. Han, T.W. Or, Z.P. Lu, and J. Heberlein, "Rapid Synthesis of Diamond by Counter-Flow Liquid Injection into an Atmospheric Pressure Plasma Jet," Diamond and Related Materials, 1 (1992), 127.

4. E. Pfender, "Heat Transfer from Thermal Plasmas to Neighboring Walls or Electrodes", Pure and Appl. Chem., 48 (1976), 199.

5. T.W. Petrie and E. Pfender, "The Effect of Ionization on Heat Transfer to Wires Immersed in a Highly Thermally Ionized Plasma," Warme- und Stoffubertragung, 5(2), (1972), 85.

6. T.N. Meyer and E. Pfender, "Experimental and Analytical Aspects of Plasma Heat Transfer," Warme- und Stoffubertragung, 6(1), (1973), 25.

7. T.W. Petrie, "The Effect of Ionization on Heat Transfer to Wires Immersed in an Arc Plasma" (Ph.D. thesis, University of Minnesota, 1969).

8. T.N. Meyer, "Effects of Applied Voltage and Surface Chemistry on the Heat Flux to a Probe Immersed in an Arc Plasma" (Ph.D. thesis, University of Minnesota, 1971).

9. J.A. Fay and N.H. Kemp, "Theory of Heat Transfer to a Shock-Tube-End-Wall from an Ionized Gas," J. Fluid Mech., 21 (1965) 659.

10. D.M. Chen, and E. Pfender, " Two-Temperature Modeling of the Anode Contraction Region of High-Intensity Arcs" IEEE Trans. Plasma Sci., PS-9(4), (1981) 265.

11. E. Leveroni, A.M. Rahal, and E. Pfender, "Electron Temperature Measurements in the Near-Wall Region of Wall-Stabilized Arcs," (Paper presented at Proc. ISPC-8, Tokyo, Japan, 1, 1982) 346.

12. E. Pfender, "Particle Behavior in Thermal Plasmas", Plasma Chem. and Plasma Proc., 9 (1) Supplement (1989), 167S.

13. E. Pfender, "Heat and Momentum Transfer to Particles in Thermal Plasma Flows", Pure and Appl. Chem., 57 (9), (1985) 1179.

14. P.J. Shayler and M.T.C. Fang, "The Transport and Thermodynamic Properties of a Copper-Nitrogen Mixture",J. Phys. D: Appl. Phys., 10 (1977) 1659.

15. J. Mostaghimi-Tehrani, and E. Pfender, " Effects of Metallic Vapor on the Properties of an Argon Arc Plasma", Plasma Chem. and Plasma Proc., 4 (2), (1984), 129.

16. R.M. Young, and E. Pfender, " Nusselt Number Correlations for Heat Transfer to Small Spheres in Thermal Plasma Flows", Plasma Chem. and Plasma Proc., 7 (2), (1987), 211.

17. G.R. Chludzinski, R.H. Kadlec, and S.W. Churchill, "Energy Transfer to Probes in Argon Nitrogen Plasmas" (Paper presented at A.I.Ch.E.I. Chem. E. Symposium Series No. 2, Chemical Engineering under Extreme Conditions, Inst. Chem. Eng., London, England, 1965), pp. 93-98.

18. M. Capitelli, F. Cramarossa, L. Triolo, and E.Molinari, "Decomposition of Al_2O_3 Particles Injected into Argon-Nitrogen Induction Plasma of I Atmosphere," Combust. Flame, 15 (1970) 23.

19. Xi Chen, and E. Pfender, "Heat Transfer to a Single Particle Exposed to a Thermal Plasma", Plasma Chem. and Plasma Proc., 2 (2), (1982), 185.

20. Y.C. Lee, "Trajectories and Heating of Particles Injected into a Thermal Plasma" (M.S. thesis, University of Minnesota, 1982).

21. M.I. Boulos, "Heating of Powders in the Fire-Ball of an Induction plasma", IEEE Trans. Plasma Sci., 4 (1978) 93.

22. T. Yoshida, and K. Akashi, "Particle Heating in a Radio-Frequency Plasma Torch", J. Phys. D: Appl. Phys., 48 (1977) 2252.

23. F.J. Harvey, and T.N. Meyer, "A Model of Liquid Metal Droplet Vaporization in Arc Heated Gas Streams", Metallurgical Trans., B 9B (1978) 615.

24. Xi Chen, and E. Pfender, "Effect of the Knudsen Number on Heat Transfer to a Particle Immersed into a Thermal Plasma", Plasma Chem. and Plasma Proc., 3 (1), (1983), 97.

25. Xi Chen, Y.C. Lee, and E. Pfender, (Proc. of the 6th International Symposium on Plasma Chemistry, 1, (1983) 51).

26. E. Leveroni, and E. Pfender, "A Unified Approach to Plasma-Particle Heat Transfer Under Non-Continuum and Non-Equilibrium Conditions", Int. J. Heat Mass Transfer, 33 (7), (1990), 1497.

27. F.S. Sherman, "A survey of Experimental results and methods for the transition regime in rarefied gas dynamics," In Rarefied Gas Dynamics, ed. J.A. Laurmann (New York, NY: Vol. II Academic Press, 1963).

28. J.A. Thornton, "Comparison of Theory and Experiment for Ion Collection by Spherical and Cylindrical Probes in a Collisional Plasma," AIAA J., 9 (1971), 342.

29. Y.C. Lee, "Modeling Work in Thermal Plasma Processing" (Ph.D. thesis, University of Minnesota, 1984).

30. H. Maecker, "Plasmastroemungen in Lichtboegen Infolge Eigenmagnetischer Kompression," Z. Phys., 141 (1955), 198.

31. H.A. Dinulescu and E. Pfender, "Analysis of the Anode Boundary Layer of High Intensity Arcs," J. Appl. Phys., 51(6), (1980) 3149.

32. N.A. Sanders and E. Pfender, "Measurements of Anode Falls and Anode Heat Transfer in Atmospheric Pressure High Intensity Arcs," J. Appl. Phys., 55(3), (1984) 714.

33. E. Leveroni and E. Pfender, "Investigation of the Anode Boundary Layer of Free-Burning, High Intensity Arcs," (Proc. of the ASME Welding and Joining Processes, Winter Meeting 1991).

34. R.C. Eberhart and R.A. Seban " The Energy Balance for a High Current Argon Arc," Internat. J. Heat Mass Transfer, 9 (1966) 939.

35. J.L. Smith and E. Pfender, "Determination of Local Anode Heat Fluxes in High Intensity, Thermal Arcs, " IEEE Transact. Power App. Syst., Vol. PAS-95 (2), (1976) 704 .

Modeling the Behavior of a Commercial Plasma Torch

with Turbulent, Swirling Flow

R. Westhoff*, J. Szekely

Department of Materials Science and Engineering
Massachusetts Institute of Technology
77 Mass. Ave., Cambridge MA 02139

ABSTRACT

The effect of swirling flow in a commercial plasma torch is represented using a mathematical model of fluid flow, heat transfer and electromagnetic phenomena, including turbulent flow in the plasma plume. The calculated data include axial and swirl velocities in the torch, temperature contours, mass flow, current density and body force vectors and current-voltage characteristics of the torch. A comparison is made with experimental temperature and velocity measurements from the literature, and while the comparison is not entirely satisfactory, the model is able to offer interpretations of some of the trends in the data. Specifically, the model shows how the axial velocity in the plume may be nearly constant or decrease with increasing torch flow rates due to swirl flow.

1. INTRODUCTION

The basic fundamental understanding of plasma torches has been significantly improved in the past decade through the contributions of mathematical modeling. Recent advances include the modeling of the fully elliptic nature the plume, swirling flow, the temperature dependent properties of the plasma and the intermixing of different gases[1-4]. More recently, a fully coupled representation of the region inside the plasma torch has been developed based on fundamental principles[5-6]. This work represented a significant advancement over previous investigations because it allowed the model to be directly applied to nearly any torch without experimental velocity or temperature data required from an experiment. The work, however, was limited to laminar cases for which specific experimental measurements had been made for the purpose of model verification. This paper describes the extension of the model to torches for which the plume becomes turbulent. These torches include most systems of practical interest for instance, in plasma spraying. The torch to be modeled is the Metco 7MB torch, shown schematically in Figure 1, for which experimental data has been presented in a paper by Capetti et al[7]. Table I summarizes the conditions used in the experiment .

* Currently at The Aerospace Corp., M4-906, P.O. Box 92957, Los Angeles, CA. 90009-2957

Plasma Synthesis and Processing of Materials
Edited by Kamleshwar Upadhya
The Minerals, Metals & Materials Society ©1993

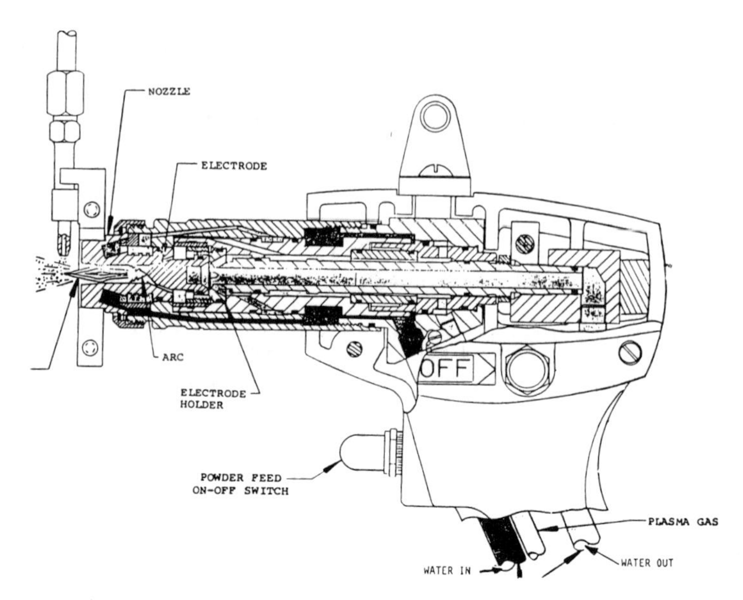

Figure 1. Cut-away view of the Metco 7mb plasma torch (courtesy Metco Corp., Westbury, NY)

Table I. Averaged conditions for the Metco torch experiments of Capetti and Pfender[7].

I (A)	Q (l/min)	V (volts)	Power (kW)	Efficiency (%)	Case # *
450	23.6	24.0	10.842	48.2	12
450	35.4	25.5	11.478	50.9	10
450	47.2	27.3	12.265	55.4	11
600	23.6	25.5	15.220	46.4	9
600	35.4	26.8	16.014	53.0	7
600	47.2	28.1	16.813	56.4	8

*These cases are labelled to be consistent with those presented in Ref. (4).

2. STATEMENT OF THE PROBLEM

Since the development of the mathematical model is based on the equations governing transport phenomena in the torch, a brief discussion of the physics follows. The direct comparison of the model to experimental measurements also requires that the boundary conditions accurately represent the actual torch inlet conditions especially with regard to swirl. To facilitate this, some details of the torch itself will be given in this section.

2.1. Description of the phenomena

The cutaway view of the plasma torch shown in Figure 1 includes a sketch of the arc which passes through the gas between the conical electrode (cathode) and the nozzle (anode). The current flow through the arc leads to Joule heating which ionizes the gas. This rapid transition to the plasma state is accompanied by expansion and increasing gas velocity. In addition, the arc current interacts with its own induced magnetic field, and the resulting Lorentz forces also accelerate the gas causing the so-called "cathode jet". The fluid dynamic, heat flow, and electromagnetic effects all interact with the torch geometry to produce its natural voltage-current versus flow rate characteristics.

For the torch in this section the injection is done in the upstream region through two tangentially directed holes 1.6 mm in diameter. The annulus upstream of the arc has an approximate inside diameter of 7 mm and and outer diameter of 8 mm, which gives a cross-sectional area of 1.178×10^{-5} m^2. The resulting values of the average velocities and radii required to quantify the swirl are those described in previous work[5]: $W = 99$ m/s at $r_{inj} = 0.0075$ m, $U = 33.4$ m/s and $R_0 = 0.0035$ m, so that

$$Sw = \frac{2}{3}\frac{W\, r_{inj}}{U\, R_0} = 4.0 \qquad (1)$$

Where W is the characteristic azimuthal velocity in for the flow exiting the tangential holes, r_{inj} is the radius at which injection takes place, U is the characteristic axial velocity just upstream of the arc, and R_0 is the characteristic nozzle radius. The type of swirl introduced is assumed to be that of solid body rotation at the inlet boundary.

In the mathematical statement of this physical picture, the fluid flow phenomena are described by the steady-state Navier-Stokes equations with due allowance for the temperature dependence of the gas density, viscosity and for the electromagnetic forces. Turbulent flow is represented using the standard K-ε model adapted for swirling flow[1]. Heat transfer is represented by the convective heat flow equation with allowances for Joule heating in the arc, thermal transport due to electron drift, thermal radiation, and energy change due to pressure variations. The electrodynamic equations include current continuity in terms of the electric potential equation and Ampere's Law which gives the self induced magnetic field. The cathode spot current density is assumed and the water cooled copper anode is considered to be an isopotential surface.

2.2. Assumptions used in the model

The assumptions used in the analysis have been described in earlier works. The equations describing the phenomena are written in their axially symmetric, steady-state form. The plasma is represented as a gas in Local Thermodynamic Equilibrium (LTE), a useful simplification which has been questioned,[8-9] yet still provides an approximate basis for developing the model. The plasma is assumed to be optically thin and the heating effects of viscous dissipation, compressibility effects and buoyancy forces due to gravity are neglected. Finally, the cathode tip is assumed to be flat rather than pointed.

3. MODEL DESCRIPTION

The model developed for the cases involving turbulence is essentially a combination of the turbulent plume model of earlier works[1,4] with the model of the phenomena inside the plasma torch[5-6]. For the sake of completeness, the full model will be presented here in summary form. The equations which describe the time averaged behavior of the system include the following:

3.1 Governing Equations

The conservation equations for mass, momentum, enthalpy, turbulent kinetic energy, turbulent energy dissipation rate and electric potential may all be cast in a similar form:

$$\frac{1}{r}\left[\frac{\partial}{\partial z}(\rho r u \phi) + \frac{\partial}{\partial r}(\rho r v \phi) - \frac{\partial}{\partial z}\left(r\Gamma_\phi \frac{\partial \phi}{\partial z}\right) - \frac{\partial}{\partial r}\left(r\Gamma_\phi \frac{\partial \phi}{\partial r}\right)\right] = S_\phi \qquad (2)$$

where ϕ represents each of the variables to be solved. The first two terms represent convection of a given property while the second two represent diffusion. The variables and the appropriate source terms are given in Table II where,

$$G_K = \mu_t\left\{2\left[\left(\frac{\partial u}{\partial z}\right)^2 + \left(\frac{\partial v}{\partial r}\right)^2 + \left(\frac{v}{r}\right)^2\right] + \left[\left(\frac{\partial w}{\partial z}\right)^2 + \left(\frac{\partial u}{\partial r} + \frac{\partial v}{\partial z}\right)^2 + \left(\frac{\partial w}{\partial r} - \frac{w}{r}\right)^2\right]\right\}$$

and the turbulent viscosity is given by ·

$$\mu_t = \frac{C_D \rho K^2}{e}$$

In these equations z, u; r, v and θ, w are the coordinates and velocities in the axial, radial and azimuthal directions respectively and ρ is the plasma density. P denotes the pressure, μ_e is the effective viscosity, j_z and j_r are the axial and radial components of the current density respectively, and B_θ is the azimuthal magnetic flux density. The terms $j_r B_\theta$, and $j_z B_\theta$ are the electromagnetic or **J X B** forces, where **J** is the current density vector, and **B** is the magnetic flux density vector.

Table II. Source terms and exchange coefficients for the general equation for ϕ, Eqn. (2)

ϕ	Γ_ϕ	S_ϕ
u	$\mu_e = \mu_l + \mu_t$	$-\frac{\partial P}{\partial z} + \frac{1}{r}\frac{\partial}{\partial r}\left(r\mu_e \frac{\partial u}{\partial z}\right) + \frac{\partial}{\partial z}\left(\mu_e \frac{\partial u}{\partial z}\right) + j_r B_\theta$
v	μ_e	$-\frac{\partial P}{\partial r} - \frac{2\mu_e v}{r^2} + \frac{\rho w^2}{r} + \frac{1}{r}\frac{\partial}{\partial r}\left(r\mu_e \frac{\partial v}{\partial r}\right) + \frac{\partial}{\partial z}\left(\mu_e \frac{\partial u}{\partial r}\right) - j_z B_\theta$
rw	μ_e	$-\frac{2}{r}\frac{\partial(\mu_e r w)}{\partial r}$
K	$\frac{\mu_e}{\sigma_K}$	$G_K - \rho\varepsilon$
ε	$\frac{\mu_e}{\sigma_\varepsilon}$	$\frac{\varepsilon}{K}(C_1 G_K - C_2 \rho\varepsilon)$
h	$\frac{\mu_e}{\sigma_h}$	$\frac{j_z^2 + j_r^2}{\sigma} + \frac{5}{2}\frac{k_b}{e}\left(\frac{j_z}{C_p}\frac{\partial h}{\partial z} + \frac{j_r}{C_p}\frac{\partial h}{\partial r}\right) - S_R + u\frac{\partial P}{\partial z} + v\frac{\partial P}{\partial r}$
V	σ	$-\frac{1}{r}\left[\frac{\partial}{\partial z}(\rho r u V) + \frac{\partial}{\partial r}(\rho r v V)\right]$

The additional source terms in the energy equation represent the transport of enthalpy by joule heating, electron drift, radiation losses, and finally a term which accounts for energy changes due to pressure variations. Symbols used in the above include: C_p , heat capacity, σ, electrical conductivity, h, specific enthalpy, k_b, Boltzmann's constant, e, the electric charge, and S_R, the volumetric radiative loss term. Also, K is the turbulent kinetic energy, ε is the turbulent energy dissipation and σ_h, σ_K, σ_ε are the turbulent Prandtl numbers for h, K and ε respectively. The constants for the K–ε model are the values recommended by Pun and Spalding[18] and are given in Table III.

Table III. Constants used in the K–ε turbulence model

C_1	1.43	σ_K	1.0
C_2	1.92	σ_ε	1.3
C_D	0.09	σ_h	0.9

The plasma thermodynamic and transport properties are taken from Liu[10] and Devoto[11] respectively and the radiation source is taken from Evans and Tankin.[12] The electrical conductivity is modified using an exponential expression below a certain cutoff temperature. This is more fully described here.

As suggested by Scott et al.[13] the electrical conductivity has been modified below a certain cutoff value using the expression $\sigma = A\ exp(T/B)$. In the expression used in Reference (13), A = 20 and B = 2000. These are the values used in previous work by the authors[5-6]. However, a parametric study indicated that more realistic values might be chosen to better represent the voltage versus flow rate behavior of the system. The constants chosen were those for which the theoretical current versus flow rate relationship (shown in Figure 7) would have the same slope as that demonstrated experimentally. For this section, the values of the constants have been modified to A = 300 and B = 4364.42. Three relations of the type chosen, together with the LTE values are plotted in Figure 3. In addition, the cathode current density used has been increased to $j_c = 6.5\text{x}10^7\ A/m^2$ based on experimental observations which indicate that higher values than previously used may be more representative of the conditions inside plasma torches[14].

Casting the Equations

It is important to note here that because of the transitional nature of the phenomena, care is required in casting the equations in a form which most accurately represents the system. In a large number of plasma torches, because of the high viscosity of the plasma, the flow within the torch will remain essentially laminar. This has been illustrated by high speed photographs and shadowgraphs which indicate laminar flow exiting the torch with a rather sudden transition to turbulent flow[15]. In this version of the model, this behavior is represented by assuming laminar flow ($\mu_e = \mu_{lam}$) inside the torch while fully turbulent flow is assumed at the nozzle exit.

3.2. Auxiliary Equations

If the current distribution is axi-symmetric, the self induced magnetic field may be calculated by the following relation derived from Amperes law:

$$B_\theta = \frac{\mu_o}{r} \int_0^r j_z\ \zeta\ d\zeta \qquad (3)$$

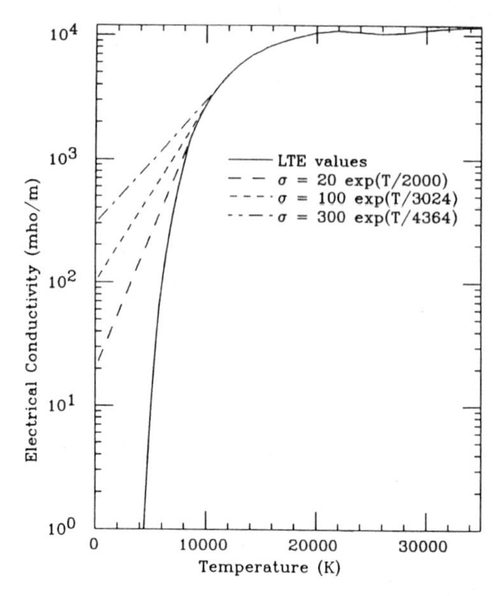

Figure 3. Temperature dependence of the electrical conductity of argon showing the assumptions used at low temperature and the effect of the constants used in the exponential expression.

where μ_o is the magnetic permeability of free space, and ζ is a dummy variable of integration.

The current density is calculated from the definition of electric potential:

$$\mathbf{J} = -\sigma \nabla V \tag{4}$$

3.3. Boundary Conditions:

The integration region is sketched in Figure 2, and the corresponding boundary conditions are given in Table IV.

These conditions specify zero velocities at all solid boundaries, and zero fluxes at the axis of symmetry. Constant temperatures of all solid boundaries and at the outer entrainment boundary are assumed. Additionally, zero gradients are assumed at the downstream boundary, and zero currents are assumed on all boundaries except on the cathode (where a current density is specified) and at the anode (where a constant electric potential is assumed). The anode and cathode surfaces may require special treatment since deviations from LTE occur in these regions.

Inlet boundary

At the inlet boundary (line DE of Figure 2) the axial velocity is assumed to have a flat profile. The radial velocity is assumed to be zero. Additionally, it is important to be able to quantify the swirl at the inlet boundary. The values given in the first section may be used to estimate swirl number. This is done in the same way as in earlier work[5-6].

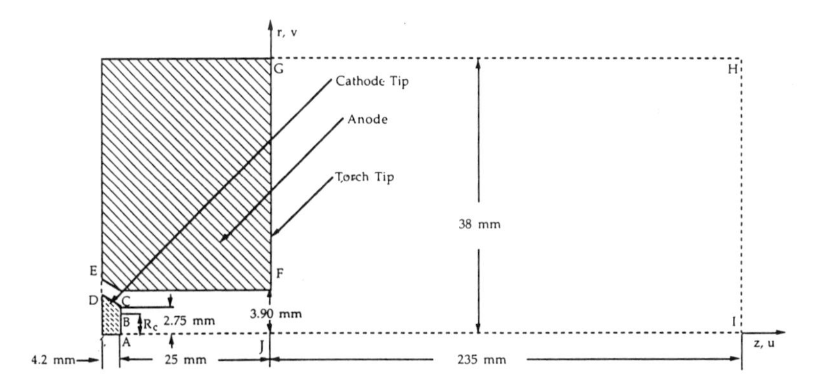

Figure 2. Computational domain used in the study of the Metco torch.

Table IV. Boundary Conditions used in the model with reference to Figure 2.

	u	v	rw	h	V	K	ε
AB	w.f	w.f.	w.f.	$T = 3000$ $Q_c = j_c V_c$	$j_c = \frac{I}{\pi R_c^2}$	-	-
BC	w.f	w.f.	w.f.	$T = 3000$	$\frac{\partial V}{\partial z} = 0$	-	-
CD	w.f	w.f.	0	$T = 3000$	$\frac{\partial V}{\partial n} = 0$	-	-
DE	$u=u(r)$	0	$rw(r)$	$T = 1000$	$\frac{\partial V}{\partial z} = 0$	-	-
EF	w.f	w.f.	0	$T = 1000$	$\frac{\partial V}{\partial n} = 0$	-	-
FG	w.f	w.f.	w.f.	$T = 700$	-	$\frac{\partial K}{\partial z} = 0$	$\varepsilon = 0$
FJ	-	-	-	-	$\frac{\partial V}{\partial z} = 0$	$K = .005u^2$	$\varepsilon = \frac{C_D K^{1.5}}{.03\, r_n}$
GH	$\frac{\partial u}{\partial r} = 0$	$\frac{\partial \rho v}{\partial r} = 0$	0	$T = 300$	-	$K \approx 0$	$\varepsilon \approx 0$
HI	$\frac{\partial u}{\partial z} = 0$	0	0	$\frac{\partial h}{\partial z} = 0$	-	$\frac{\partial K}{\partial z} = 0$	$\frac{\partial \varepsilon}{\partial z} = 0$
IA	$\frac{\partial u}{\partial r} = 0$	0	0	$\frac{\partial h}{\partial r} = 0$	$\frac{\partial V}{\partial r} = 0$	$\frac{\partial K}{\partial r} = 0$	$\frac{\partial \varepsilon}{\partial r} = 0$

Electrodes

The regions near the electrodes are represented as follows. The contribution of the cathode fall is given approximately by the "free fall" type of expression for the cathode fall voltage, V_c as was done by McKelliget and Szekely(16) for a free-burning arc:

$$V_c = \frac{5}{2}\frac{k_b T_e}{e} \quad (5)$$

In this expression, V_c is the cathode fall voltage, and T_e is the electron temperature, which is approximated as the maximum plasma temperature in the column adjacent to the cathode (~20000 K), so that $V_c \sim 4.3$ V. In Table IV, the torch current is denoted by I, and Q_c is a positive source to the plasma column at the cathode boundary which approximates the energy used in the cathode boundary layer to ionize the plasma.

The boundary condition for electric potential is approximated assuming that the cathode current density, j_c which is emitted from the cathode normal to the surface is constant inside the cathode spot radius, R_c and is zero outside.

Nozzle Exit

At the nozzle exit, where transition from laminar to turbulent flow is assumed to occur, the values of K and ε must be set. The default values for laminar flow would be zero, but this was seen to grossly underestimate the turbulent mixing which occurs in the plume. For this reason, the values were calculated using the expressions in Table IV which represent turbulent flow[18]. This leads to somewhat of a paradox at the nozzle exit, namely, assuming laminar flow, but specifying values of K and ε (for use in the plume) which correspond to turbulent flow. If this is not done, however, the turbulence develops much too slowly to represent the plume behavior.

3.4 Solution Method

The solution is done using a finite volume technique embodied within a modified and extended version of the code developed by Dilawari et al.[17] based on the 2/E/FIX code of Pun and Spalding[18]. The code, which has been modified to include the electromagnetic phenomena, was executed in a mode which includes the solution of the turbulence equations in the plume region. The grid system was previously found to give sufficient numerical accuracy, and uses a 67X40 mesh, with 33X15 grids inside the torch and 34X40 grids outside. The solution requires about 4.5 hours of CPU time on a Vaxstation 3100 for three thousand iterations.

4. RESULTS OF CALCULATIONS

In the following we shall present some results of calculations performed using the experimental conditions listed in Table I which were taken by Capetti and Pfender[7]. Table V summarizes some of the results from the calculations.

Table V. Summary of calculated results for the Metco torch.

Run No.	V_{arc} (calc.) (V)	Exit Re No.	Exit Swirl No.	Net Power (W)	Axis Temp. in torch (K)	Axis Temp. at exit (K)	Axis Velocity in torch (m/s)	Axis Velocity at exit (m/s)
12	22.2	416	0.236	4993	25600	14030	1159	563
10	23.8	621	0.315	6332	24970	14160	1661	545
11	25.1	849	0.381	7384	24750	14180	1297	492
9	23.0	439	0.197	6580	26580	14770	1300	748
7	24.8	632	0.257	8468	25910	14920	1853	744
8	26.3	830	0.305	9938	25410	14890	2406	625

Figure 4 (a-b) shows a set of representative temperature contours and velocity vectors in the system (for case 12, 450 amps, 23.6 liters/min). The maximum temperature is higher than in previous references (5-6) due to the higher cathode current density. The resulting maximum velocity of 1159 m/s is also much higher than in those works because of the higher flow rate and current density.

Figure 5 (a-b) shows the mass flow (ρu product) vectors and the contours of the swirl velocity in the system. The mass flow vectors show a maximum near the walls because of the high mass density in those lower temperature regions. This is in agreement with the results shown by previous investigators(19-20). The swirl isocontours show a maximum (901 m/s) on the center line as a result of the combination of decreased density and the conservation of angular momentum (swirl velocity tends to increase as 1/r).

Figure 6 (a-b) shows the electric current density vectors and the resulting body forces in the flow. The maximum current density is 6.3×10^7 A/m^2 which occurs near the cathode tip. The body forces result from the interaction of the current density with the self-induced magnetic field according to the right-hand rule, (maximum $= 2.5 \times 10^6$ N/m^3). As a result, the body forces act toward the center line of the system which tends to increase the pressure on the axis away from the cathode tip. These forces lead to the well known "cathode jet" effect.

Figure 7 shows a comparison of the calculated and experimentally measured voltage characteristics of the torch as a function of the gas flow rate for the two different currents. The calculations represent the experiments rather well, however, the calculation both neglects to include any allowance for the anode fall and includes the assumption about the plasma electrical conductivity, which together could lead to the observed discrepancy of some 2 volts.

Figures 8-11 (a-b) illustrate the behavior of the plasma plume as it exits the nozzle of the torch. Figures 8 and 10 show the comparison between the measured and calculated profiles of the temperature and velocity on the center line of the plume versus axial position for two different current levels and three different flow rates. The discrepancies are partly due to the limitations of switching from a laminar model inside the torch to a fully turbulent model at the torch exit.

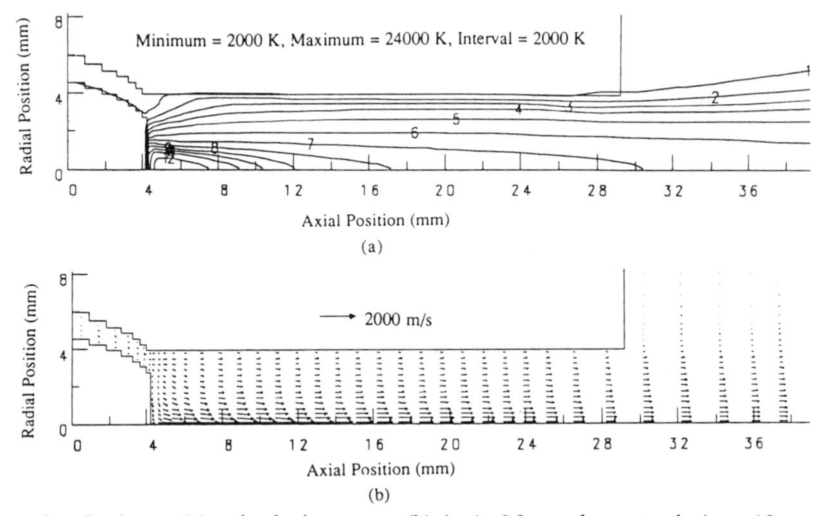

Figure 4. Isotherms (a) and velocity vectors (b) in the Metco plasma torch, (case 12, 450 A, 23.6 lit./min. of argon, Sw=3.0).

While the comparison is not as good as may be obtained by representing flow with a "plume only" model,[21] using the methods of references (1) and (4), at least one of the trends shown may prove to be more realistic. The observation in question is that with an increase in flow rate, the velocities on the axis of the plume do not change appreciably, or may even decrease[7]. It was suggested in reference (7) that this was caused by the decrease in temperature and corresponding density increase with increasing flow rate.

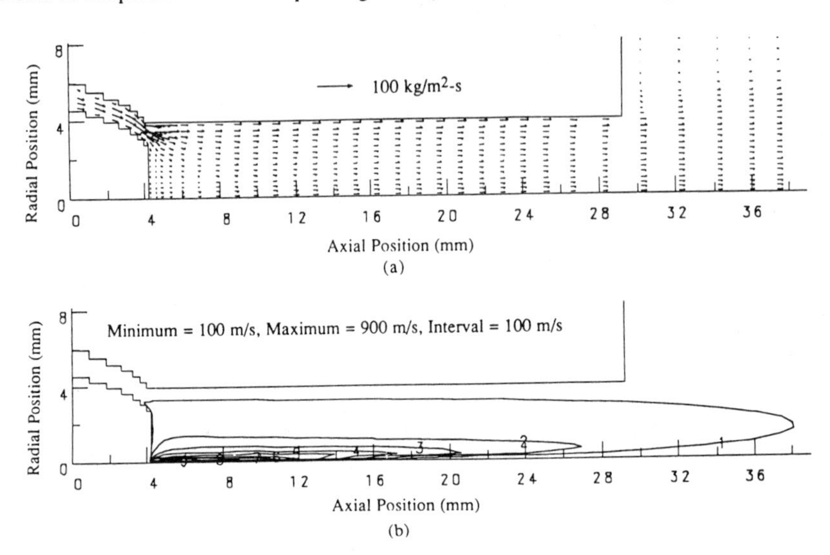

Figure 5. Mass flow (ρu product) vectors (a) and contours of swirl velocity (b) in the Metco plasma torch, (case 12, 450 A, 23.6 lit./min. of argon, Sw=3.0).

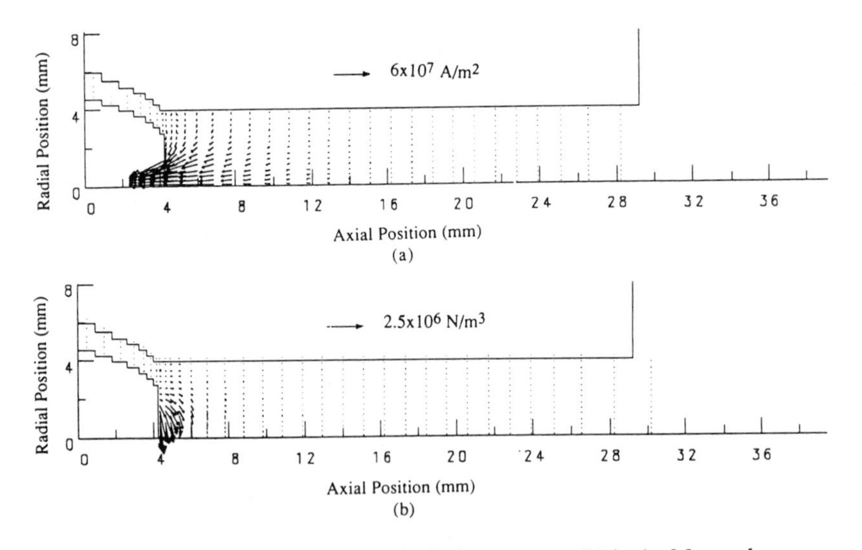

Figure 6. Current density vectors (a) and body force vectors (b) in the Metco plasma torch, (case 12, 450 A, 23.6 lit./min. of argon, Sw=3.0).

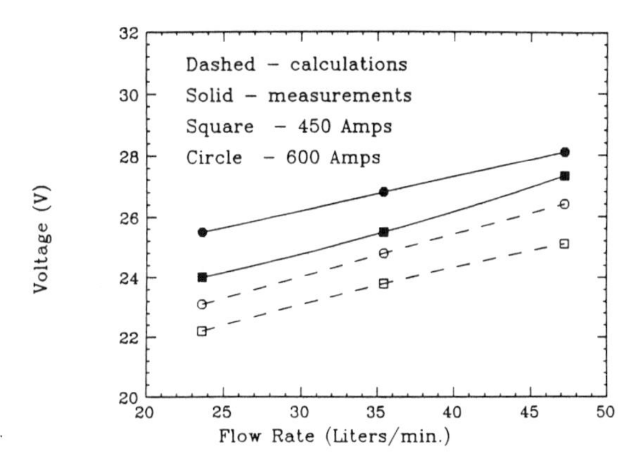

Figure 7. Calculated and experimentally measured current-voltage characteristics of the plasma torch used by Capetti and Pfender[24], calculated (open symbols) and measured (solid symbols).

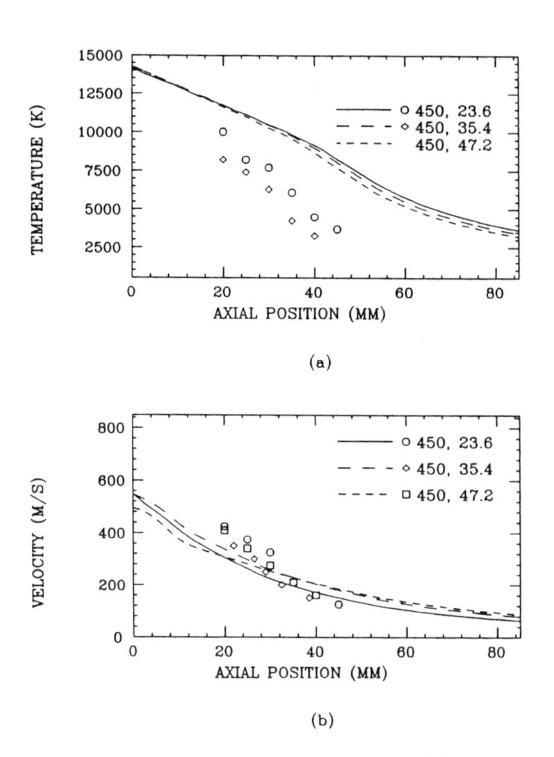

Figure 8. Comparison between experimentally measured[24] (symbols) and theoretically predicted (lines) axial profiles of temperature (a) and axial velocity (b) in the plume of the Metco torch for the 450 amp cases.

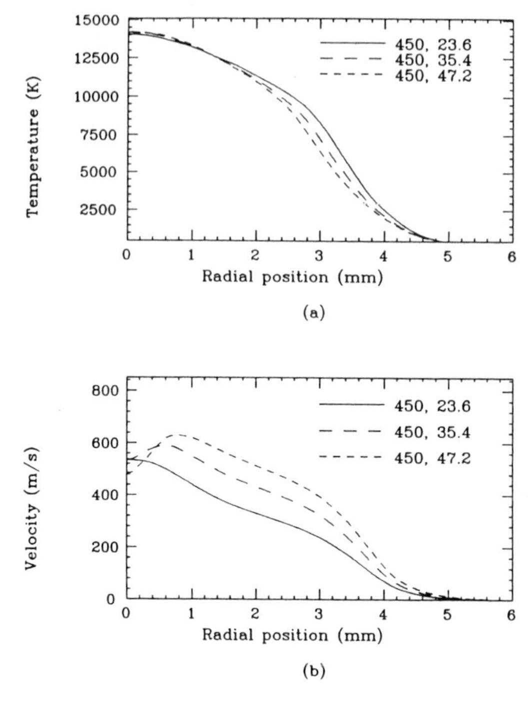

Figure 9 Theoretically predicted radial profiles of temperature (a) and axial velocity (b) at a position 1 mm from the nozzle exit for the 450 amp cases.

This, however, is contrary to what may be seen from a simple mass balance as summarized in Table VI: The average enthalpy at the torch exit can be derived from the net power and mass flow. The average temperature and density can then be derived from gas property tables. From this, the average velocity may be derived, which is seen to increase with increasing flow. This illustrates that the density increase alone cannot account for the velocity effect observed in reference (7).

Table VI. Summary of the overall heat and mass balances performed on the Metco torch

Case #	Net power (W)	Mass Flow (kg/s)	Average Enthalpy (J/kg)	Average Temperature (K)	Average Density (kg/m^3)	Average Velocity (m/s)
12	5226	7.01e-4	7.45e6	10575	0.0460	312
10	5842	1.05e-3	5.56e6	9454	0.0515	418
11	6795	1.40e-3	4.85e6	8794	0.0554	519
9	7062	7.01e-4	1.01e7	11466	0.0425	339
7	8502	1.05e-3	8.10e6	10876	0.0448	481
8	605	1.40e-3	6.77e6	10255	0.0475	605

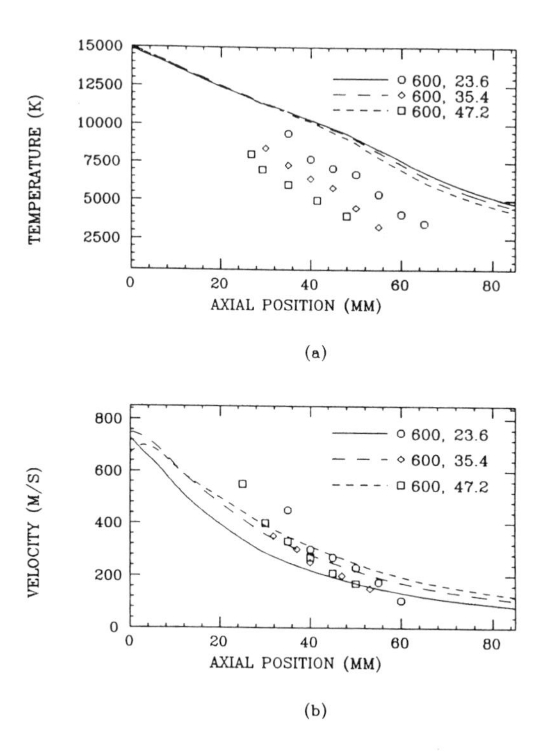

Figure 10. Comparison between experimentally measured[24] (symbols) and theoretically predicted (lines) axial profiles of temperature (a) and axial velocity (b) in the plume of the Metco torch for the 600 amp cases.

The model, however, suggests an explanation for this trend. It appears that the high flow case may be more "swirl dominated" at the nozzle exit, which leads to a wider jet spreading and thus comparable or lower velocities on the center line.

This effect is further illustrated in Figures 9 (a-b) and 11 (a-b) which show the radial profiles of temperature and velocity at a position 1 mm from the nozzle exit. The figures show that while the temperature profile is relatively unchanged as the flow rate is increased, the velocity profile may change significantly, even exhibiting a bi-modal distribution at higher flow rates. This occurs because the larger flow of gas receives a relatively smaller increase in axial momentum due to expansion forces than the smaller gas flow, resulting in a higher swirl number at the nozzle exit as seen in Table V. The enhanced spread of the jet may lead to lower velocities in the plume, which is seen for example in Figure 8 over the section of the plume nearest the nozzle exit

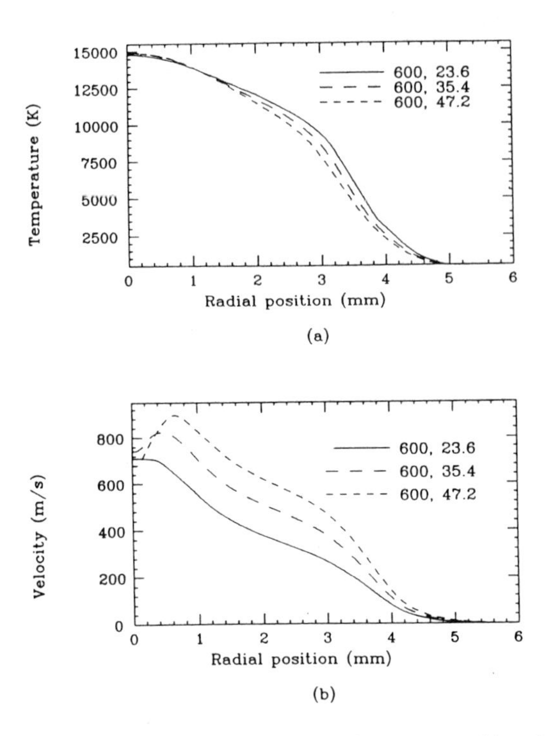

Figure 11. Theoretically predicted radial profiles of temperature (a) and axial velocity (b) at a position 1 mm from the nozzle exit for the 600 amp cases.

5. CONCLUSIONS

This paper has illustrated the applicability of the model to a plasma torch operating at high flows like those which are usually used in practical applications. The model shows some significant limitations, tending to over-predict the temperatures and under predict the velocities. It is felt that these shortcomings stem from a number of simplifications and assumptions, including the swirl number at the entrance of the torch, the assumption of LTE, the assumption of axial symmetry, which necessitates the use of an artificially high electrical conductivity, and finally the simplifications made regarding the transition from laminar flow in the torch to fully turbulent flow in the plume. In addition, greater discrepancies have been observed between experimental measurements and predictions for pure argon systems (such as modeled here) than for those in which the torch exits into an air atmosphere(21), perhaps indicating a theoretical difficulty that is masked by the intermixing with the higher heat capacity gas.

The model, however, has provided some important insights into the behavior of the torch, including an inside look into the persistence of swirl in the torch nozzle. This will be explored in some more detail in future publications.

6. ACKNOWLEDGEMENTS

The authors wish to acknowledge the support of the Department of Energy for support of this research under grant #DE-FG02-85ER13331. In addition, a great debt is owed to Prof. A.H. Dilawari for his work done in this lab which laid a unique and valuable foundation for this work.

7. REFERENCES

1. A.H. Dilawari and J. Szekely, Int. J. Heat and Mass Transfer, **30**, No. 11, 2357, (1987).
2. A.H. Dilawari, J. Szekely, J. Batdorf and C.B. Shaw, Plasma Chem. Plasma Proc., **10**, No. 2., (1990).
3. Y.P. Chyou and E. Pfender, Plasma Chem. Plasma Proc., **9**, (2), 291, (1989).
4. A.H. Dilawari, J. Szekely and R. Westhoff, Plasma Chem. Plasma Proc. **10**, (3), 501, (1990).
5. R. Westhoff and J. Szekely, J. Appl. Phys., **70**, (7), 3455-3466, (1991).
6. R. Westhoff, and J. Szekely, Proceedings of the International Symposium: "Thermal Plasma Applications in Materials and Metallurgical Processes", San Diego., 139-152, (1992).
7. A. Capetti and E. Pfender, Plasma Chem Plasma Proc., **9**, (2), 329-341, (1989).
8. K.C. Hsu and E. Pfender, J. Appl. Phys. **54**, No. 8, 4359-66, (1983).
9. D.M. Chen, K.C. Hsu and E. Pfender, Plasma Chem Plasma Proc., **1**, No. 3, 295-314, (1981).
10. C.H. Liu, Ph.D. Thesis, Department of Mech. Eng., Univ. of Minnesota, Minneapolis (1977).
11. R.S. DeVoto, Phys. Fluids, **16**, No. 5, 616, (1973).
12. D.L. Evans and R.S. Tankin, Phys.Fluids **10**, No. 6, 1137-44, (1967).
13. D.A. Scott, P. Kovitya, and G.N.Haddad, J. Appl. Phys., **66**, No. 11, 5232, (1989).
14. P. Fauchais, Comment on the author's paper made in the sessions of the TMS International Symposium: "Thermal Plasma Applications in Materials and Metallurgical Processing", San Diego, (1992)
15. J. Fincke, T.E. Repetti, S.C. Snyder, G.D. Lassahn, B.A. Detering, Proceedings of the TMS International Symposium: "Thermal Plasma Applications in Materials and Metallurgical Processing", 85-94, San Diego (1992).
16. J. McKelliget and J. Szekely, Met. Trans. A, **17A**, 1139-48, (1986).
17. A.H. Dilawari, J. Szekely and R. Westhoff, ISIJ International, **30**, No. 5, (1990).
18. W. Pun and D.B. Spalding, Report No. HTS/76/2, Heat Transfer Section, Imperial College, London, (1976)
19. V.R. Watson and E.B. Pegot, NASA TN D-4042, (1967).
20. A. Mazza and E. Pfender, Symp. Proc. Int. Symp. Plasma Chem. 6th, **1**, 41-50, (1983).
21. R. Westhoff, Ph.D. Thesis, Dept. of Materials Science and Engineering, Massachusetts Institute of Technology, (1992).

THE REVERSE-POLARITY PLASMA TORCH

Its Characteristics and Application Potentials

Dr. Salvador L. Camacho

Chief Scientist, First Mississippi Corporation

A B S T R A C T: The Reverse-Polarity (R-P) Plasma Torch is a versatile low-mass, high temperature heater. The torch converts electricity into heat via the Joule heating of a very small flowrate of gas that has been energized to the plasma state. The arc column in the plasma state is stabilized within the core of a rotating gas.

The Joule-heated arc column of the plasma torch readily heats up the rotating gas that surrounds the column to high temperatures of 2,000-5,000 C, temperatures that are not achievable by fossil fuel burners. The plasma-heated gas may then be employed to heat any material by conduction, convection, and radiation. The virtually massless heat generated in the plasma torch permits the attainment of controlled high temperatures and enthalpies, which are desirable for melting, cladding, metallizing, speroidizing, pyrolyzing, synthesizing, etc..

The R-P plasma torch is available for either transferred or non-transferred heating applications. The transferred torch is preferred for heating or melting solids, and the non-transferred torch is preferred for heating liquids and gases.

This paper will describe the R-P plasma torch, discuss some of the advantages and benefits of heating with this torch, and mention briefly some industrial applications of the R-P plasma torch.

1. INTRODUCTION

Plasma heating technology, during the last decade, has proven to be a very controllable and appropriate heat source for promoting desirable physical and chemical changes in matter. The use of the technology is expanding into many sectors of industry and commerce.

The employment of "plasma heat" in industry began in 1878 with the use of the DC Furnace developed by Sir W. Siemens.[1] Since then, plasma heating technology has been enhanced by many important developments that followed Siemens' DC Furnace. These developments include the works of: Berthelot, Moissant Heroult, Birkeland & Eide, Maecker, Schoenherr, etc..

Plasma heating technology received a timely impetus for further development in the late 1950's by the U. S. NASA Space Program. The unique heating environment created by the plasma torch was employed successfully by NASA to simulate the temperatures encountered during re-entry by space vehicles. The technology verified the effectiveness of heat shield materials in protecting spacecrafts during re-entry into earth's dense atmosphere.

The American Astronauts successfully landed on the moon and return safely to earth in the late 1960's. Following this significant world event, the industrial sectors of the United States, the European Countries, and the Commonwealth of Independent States, (the former USSR), "evaluated" plasma heating technology for various industrial applications. The evaluators confirmed the many benefits of plasma heating. And they accepted plasma heating technology with its needs for continued development, to adapt the technology to industrial processes.

Today, plasma technology is receiving considerable attention from industries around the world because of its unique heating abilities: low-mass heat delivery, high gas enthalpy, controlled high temperature, wide choice of plasma gas to accomodate different processes, and the availability of affordable electric power.

I believe that plasma heating technology has the potential to revolutionize the processes of metals and refractory production, materials fabrication, hydrocarbon gasification, environmental remediation, et cetera.

2. PLASMA ARC TORCHES[2]

2.1 There are many types, sizes, and shapes of plasma arc torches. A plasma torch today may be characterized by: (1) its unique form or shape, (2) its operating parameters, and (3) its essential requirements for operation. Let me familiarize you with today's plasma arc torches using these characteristics.

---

1. Feinman, Jerome, "Plasma Technology in Metallurgical Processing," Chapter 2, Iron & Steel Society, 1987.
2. Camacho, S. L., "Industrial-worthy Plasma Torches, State-of-the-Art," Paper No. S1-01, ISPC-8, Tokyo, Japan, 1987.

2.2 The plasma torches of a particular supplier may be characterized based on its UNIQUE FORMS or SHAPE:

A - **Tube-on-a-box.** This is characteristic of the Aerospatiale, Westinghouse, SKF, Huls, and other torches that require magnetic field rotation of the arc attachment. The box design accommodates important features for magnetic field rotation of the arc attachment, for rapid torch interchangeability, and for safety.

B - **Tubular.** This is characteristic of the Union Carbide, Retech, Paton Institute, Voest-Alpine, Tetronics, Plasma Energy Corp., Krupp-Mannesmann-Demag, and similar torches. Induction plasma torches are also tubular in shape.

C - **Pistol shaped.** This is characteristic of plasma welding and cutting torches by Metco, Plasmadyne, Ionarc, Electro-plasma, Plasma Technik, and similar torches.

2.3 A plasma torch may be characterized also by its OPERATING PARAMETERS:

A - **Operating Power.** Low power @ W = 100 kW or less, Moderate Power @ 100 kW < W < 2,000 kW, and High Power @ W = 2,000 kW and higher. [This is a suggested characterization of torch operating power.]

B - **Operating Mode.** Operable in transferred mode only, or operable in non-transferred mode only. Or the torch is field convertible to operate in either mode. In non-transferred mode of operation, the plasma columnm is either a fixed-length arc (as in the segmented electrode torch by SKF, Tioxide, Accurex, etc.), or a free-length arc (as in the Westinghouse, Aerospatiale, UCC-Retech, Plasma Energy, Tetronics, Krupp-MD, and similar torches).

C - **Operating Gas.** Operable with any plasma gas, including air and oxygen. Operable only with oxygen-free gases, e.g., tungsten-electrode torches. Or operable only with a prescribed list of suggested gases or gas mixtures.

D - **Operating Electrode.** Solid tungsten electrode; hollow electrode made of different materials--copper, silver, steel, etc.; or no electrode is required, as in the Induction Plasma Torch. The rear electrode may be connected to a positive potential (the Plasma Energy Reverse-polarity connection) or to a negative potential (the straight-polarity connection of other plasma torches.)

E - **Method of Starting.** The methods of starting a plasma torch include: (1) mechanical shorting bar, (2) high-voltage high frequency, or (3) high-voltage, single-pulse. The same plasma gas may be specified for start-up and continuous operation. Or one type of plasma gas and flowrate may be specified for the start-up, and another gas and flowrate may be specified for continuous operation.

2.4 The TORCH REQUIREMENTS may be used to characterize a plasma torch.

A - **Power Source.** Direct Current, Alternating Current @ 50 or 60 Hertz, or for Induction plasma torches, AC @ kilo-Hertz or mega-Hertz frequencies. The plasma torch may be optimized for high current, low voltage operation (as in the tungsten-tip torch) or optimized for moderate current, high voltage operation (as in the hollow-electrode torches.)

B - **Coolant.** De-ionized water or regular drinking water may be used. The plasma torch cooling water is cooled by a water-air or by a water-water heat exchanger and recycled.

3. THE REVERSE-POLARITY PLASMA ARC TORCH[3]

The R-P plasma torch may be described using the elements of characterization presented above.

The R-P plasma torch is a **tubular torch.** The tube is made of stainless steel, and the purpose of the tube is to integrate the electrodes, insulators, gas injectors, water dividers, etc., into a functional torch. See Figure No. 1.

The electrodes are hollow cylinders made of alloyed copper. The Rear Electrode serves both modes of operation--transferred and non-transferred. Another hollow cylinder is spaced co-axially about one centimeter in front of the Rear Electrode. The plasma gas is injected tangentially between the cylinders.

For transferred mode of operation, the front cylinder is short and is called the Collimator. Its function is to collimate and guide the plasma column as it rotates and exits the plasma torch to effect an external cathode attachment point.

For non-transferred mode of operation, the front hollow cylinder is longer and is called the Front Electrode. Its function in the R-P torch design is to provide the cathode attachment point of the plasma column.

The Front Electrode of the R-P plasma torch features an expanded diameter that is designed to provide a low-pressure region as the preferred surface for the cathodic arc attachment point. See Figure No. 2. The expanded diameter region allows the arc column to "extend" in length beyond the front face of the Front Electrode, exposing a part of the length of the arc column. The extended arc column beyond the Front Electrode results in: 1-higher power, derived from the higher voltage, and 2-hotter plasma flame temperature, hotter by 3,000 C or more. Figure No. 3 depicts this gain in plasma flame temperature and higher power.

The wear pattern of the Front Electrode in the R-P plasma torch is unique. The wear pattern is **axial, front-to-rear,** unlike the radial wear pattern of the front electrode in Straight-polarity plasma torches. The wear pattern in the Rear Electrode of the R-P plasma torch is **radial**, the same as in the Straight-polarity torches. See Figure No. 4.

---

3. Camacho, S. L., "The Reverse-Polarity Plasma Torch, Paper No. BV-07, ISPC-8, Tokyo, Japan, 1987.

The R-P plasma torch does not employ magnetic field for arc attachment rotation. This permits the design of a simple tubular torch that can be made for any length to accommodate any furnace design. Stainless steel tube is used for the torch shroud, and the tube may be machined to fit into an O-ring sealed flange.

The Reverse-polarity plasma arc torch is supplied as a component of a complete Plasma Heating System. See Figure No. 5. The other major components include the following:

I - **DC POWER SUPPLY**

II - **WATER-COOLING SYSTEM**

III - **PLASMA GAS SYSTEM**

IV - **WATER-GAS MANIFOLD**
[This component includes all the water and gas control and monitoring equipment that are installed for simplifying the operation of the Plasma Heating System from a Computerized Control Panel.]

V - **COMPUTERIZED CONTROL PANEL**

VI - **WATER/POWER JUNCTION BOX**
[This component integrates water and power connections to the Reverse-polarity Plasma Torch.]

The typical operating ranges of the Reverse-polarity Plasma Torch are: External Arc Length of 4-10 inches for low power and 10-40 inches for high power; Voltage/Current of 200-400 Volts/200-500 Amperes for low power and 500-1,200 Volts/400-5,000 Amperes for high power; Plasma Gas flowrate of 12-150 SCFM; and Cooling Water flowrate of 30-120 GPM. Drinking water quality is specified for cooling the R-P plasma torch electrodes; de-ionized water may be used but is not required. The range of torch Input Power is 100 kW to 5,000 kW. Electrode life is in the range of 100 hours to 1,000 hours, depending on the electrode material and the plasma gas and flowrate.

Torch input power of the transferred torch is varied by moving the plasma torch to change the arc length, changing the arc current, or both. The Input power of the non-transferred torch is varied by changing the gas flowrate, changing the arc current, or both. Figure No. 6 depicts the test results of the R-P Plasma torch with Argon and Helium plasma gases, operated at a pressure range of 220 to 625 Torr. The Arc Voltage versus Arc Length Tests were conducted at 1,200 Amperes Arc Current.

The R-P Plasma Torch was tested at 1 Atmosphere Pressure and at different Power Level, Arc Current, and with different Plasma Gas. The following data show examples of Voltage Gradient for Argon + 5% CO, Oxygen, Nitrogen, Air, and Methane.

- Argon + 5% CO, 1 kA -- E = 10 Volts/Cm
- Oxygen, 3 kA -- E = 10.5 Volts/Cm

- Nitrogen, 2 kA -- E = 10.5 Volts/Cm
- Air, 3 kA -- E = 10.5 Volts/Cm
- Methane, 1 kA -- E = 13.4 Volts/Cm

4. THE BENEFICIAL CHARACTERISTICS OF THE R-P TORCH

The R-P plasma torch is available for either mode of operation. The non-transferred mode may be used for heating gases, liquids, and solids to promote rapid physical and chemical changes. The transferred mode is recommended for fusing, smelting, and melting processes.

Almost any type of plasma gas may be used to operate the R-P plasma torch. These gases and gas mixtures have been used: Air, Oxygen, Nitrogen, Argon, Argon + CO, Helium, Hydrogen, Methane, CO, and others.

The lower (than **anode**) power drop at the **cathode** attachment point in the transferred mode of operation may be exploited for controlled-power heating, i.e., during metal deposition.

When operated in the non-transferred mode, the special design of the front electrode allows the plasma arc column to be exposed in front of the torch. The exposure of the arc column results in higher achievable temperatures in the plasma flame.

With the advent of programmable controllers and personal computers, the control of the R-P plasma torch system is greatly simplified and automated. Plasma Heating Systems that employ the R-P plasma torch are easy to operate and maintain.

5. APPLICATION POTENTIALS OF THE R-P PLASMA TORCH

5.1 HEATING MOLTEN METAL.

Molten metal in a ladle or in a tundish require a source of heat for temperature maintenance or for temperature addition. The R-P plasma torch has been demonstrated to be an effective source of heat for these applications.

Ladle heating was demonstrated at USX Lorain Works in Ohio and at Chaparral Steel in Midlothian, Texas.4 Tundish heating by the R-P plasma torch is currently part of the steelmaking process at Chaparral Steel and at other steelmills in the US, Europe, and Japan. A 3.5 mW R-P plasma torch was used at USX to maintain the temperature of molten metal in a 220-ton Ladle. See Figure No. 7. A 4.5 mW R-P plasma torch was used at Chaparral to maintain or to add temperature to molten metal in a 150-ton Ladle. See Figure No. 8.

---

4. See "The Use of Plasma Ladle Heater in Continuous Casting," pages 151-158, World Steel and Metalworking, Fachberichte Magazine, Volume 9, 1988.

The R-P plasma torch operated in transferred mode and anchored the cathode attachment point of the arc column on the molten metal. The molten metal completed the electrical circuit via a carbon rod immersed in the melt at USX or via conducting bricks that are in direct contact with the ladle at Chaparral.

The plasma torch was operated with nitrogen, and Captive Argon Bubbling (CAB) was used to stir the molten steel. Nitrogen pick-up by the molten steel was not detectable. The plasma ladle heater at USX routinely maintained the temperature of molten steel in the ladle for as long as necessary to restore an inoperative vertical continuous caster, to fix a broken crane, to prepare a teeming station, or for other reasons that accommodated the steelmaking process at Lorain. The Chaparral Steel plasma ladle heater **maintained** temperature or **raised** temperature at the rate of 1 degree C per minute.

The R-P plasma torch is employed also in Tundish Heaters. Small 1-2 mW plasma torches are used to maintain or to raise temperatures of molten metal in the tundish of Vertical and Horizontal Continuous Casters.

A distinguishing feature of the R-P plasma torch operation in ladle and tundish heating is its stability. A 1,000-volt, 4-5 kilo-Amps plasma column can be maintained for a cumulative operating time of 160 hours or more between electrode change. Other benefits of plasma ladle and tundish heating are: simple installation and start-up, versatile operation with any gas atmosphere, and no carbon or nitrogen pick-up.

5.2 RECOVERING ALUMINUM METAL FROM DROSS.

Dross is a waste by-product from the aluminum metal production process. When molten aluminum is exposed to ambient air, part of the metallic aluminum is oxidized. The oxides that are formed freeze at higher temperature than the freezing temperature of molten aluminum. And the metallic aluminum and other metal impurities freeze on the oxide, generating "aluminum dross." The dross may contain as much as 50+% aluminum metal.

The separation of the metallic aluminum from the dross has been a difficult process. Before plasma heating, the metallic aluminum could not be efficiently separated from the oxides. It was necessary to add various salts to the dross, in order to lower the temperature at which metallic aluminum is released from the oxides. The salt addition formed salt-contaminated wastes that complicated the disposal process.

The R-P plasma torch has been demonstrated successfully for the recovery of metallic aluminum from dross. The process is very simple, and no salt or other additive is required to isolate the metallic aluminum from the oxides and impurities.

A Reverse-polarity non-transferred plasma torch is used to heat the dross and effect the separation between the metallic aluminum and the aluminum oxides and impurities that are in the dross. The furnace is rotated to tumble and expose the dross to the plasma flame. Air is used as the plasma gas. And the product gas from the furnace is exhausted to the open air via a particulate filter.

In this plasma-heated process, no additives to the dross is required. The unique ability of the plasma torch to heat materials to high temperatures makes it possible to heat the high-melting point oxides of aluminum and to release the metallic aluminum from the oxides. Nearly 100 % of the metallic aluminum is recovered from the dross.

Two industrial plants for aluminum metal recovery are in operation using the R-P plasma heater. One plant is at Alcan in Jonquier, Canada, and another plant is at Plasma Processing Corporation (PPC) in West Virginia. The PPC Plant processes approximately 100 million pounds of dross per year, recovering about 50 million pounds of aluminum metal. PPC is a subsidiary of First Mississippi Corporation of Jackson, Ms.

5.3 MELTING TITANIUM SCRAPS:

Titanium metal, because of its low weight, high strength characteristics, is employed in the manufacture of rotating machinery components and this precious metal demands the careful control of temperature, metallurgy, and purity.

High temperature is required for melting Titanium. And the conventional heat source for recovery of titanium metal from scraps is Electron Beam. But E-B melting is effective only in vacuum, otherwise the Beam of heat energy is absorbed by gaseous molecules and the melting efficiency is reduced significantly. During E-B melting of the alloyed titanium metal scraps, the alloying elements of Aluminum and Vanadium are vaporized. The requirements for high temperature and for minimizing the loss of alloying elements can be satisfied by plasma heating of the scrap with the R-P plasma torch. See Figure No. 9.

The Timet Company of Nevada is using this three R-P plasma torch-heated furnace for melting scraps of titanium at near atmospheric pressure to prevent the vaporization of allowing elements. One R-P plasma torch is used to melt the incoming scraps, while another R-P plasma torch is used to maintain the temperature of the molten metal in the hearth. The third R-P plasma torch is used to maintain temperature in the withdrawal mold, in order to insure a dense ingot. This furnace is effective also in preventing high-density and low-density inclusions in the product ingot.

SUMMARY: The Reverse-Polarity plasma torch was described and its important characteristics and benefits were highlighted. The unique capability of the R-P plasma torch to generate a massless high temperature gas was exploited for: (1) heating and maintaining the temperature of liquid metals, (2) heating aluminum dross and isolating metallic aluminum for recovery, and (3) heating scraps of titanium and recovering high-density, as-melted alloyed titanium metal.

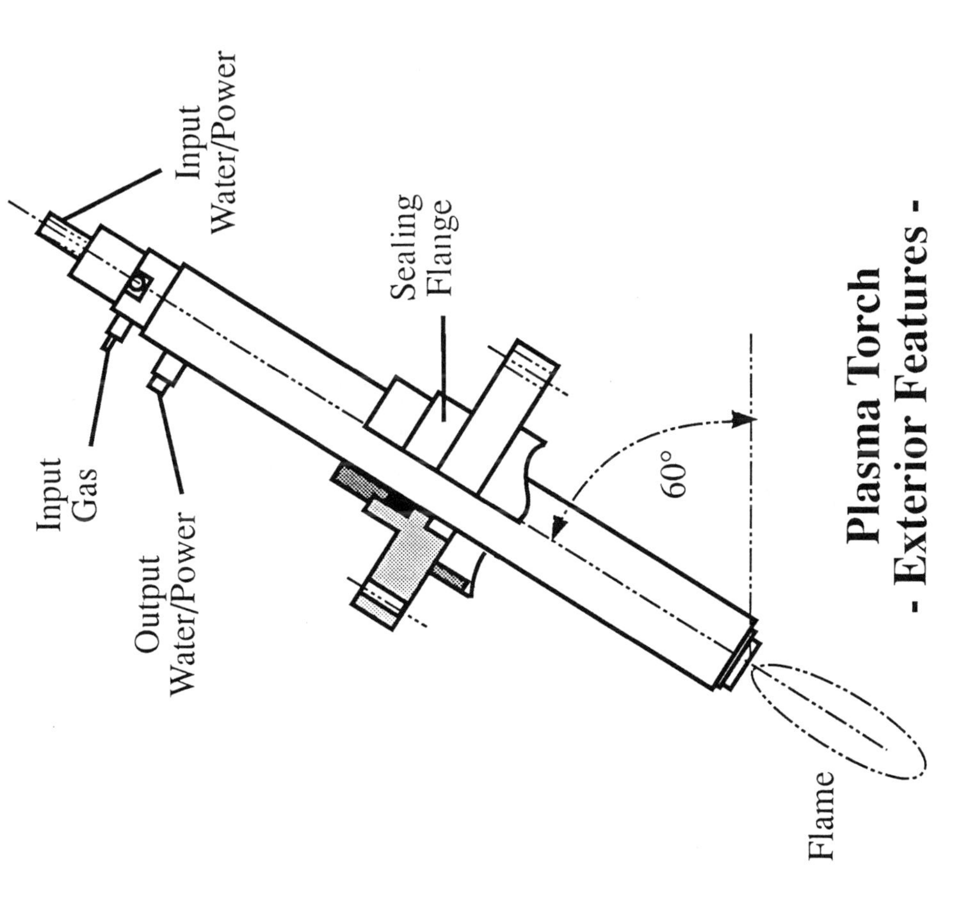
Input
Water/Power
Sealing
Flange
Input
Gas
Output
Water/Power
60°
Flame
Plasma Torch
- Exterior Features -

Figure 1.

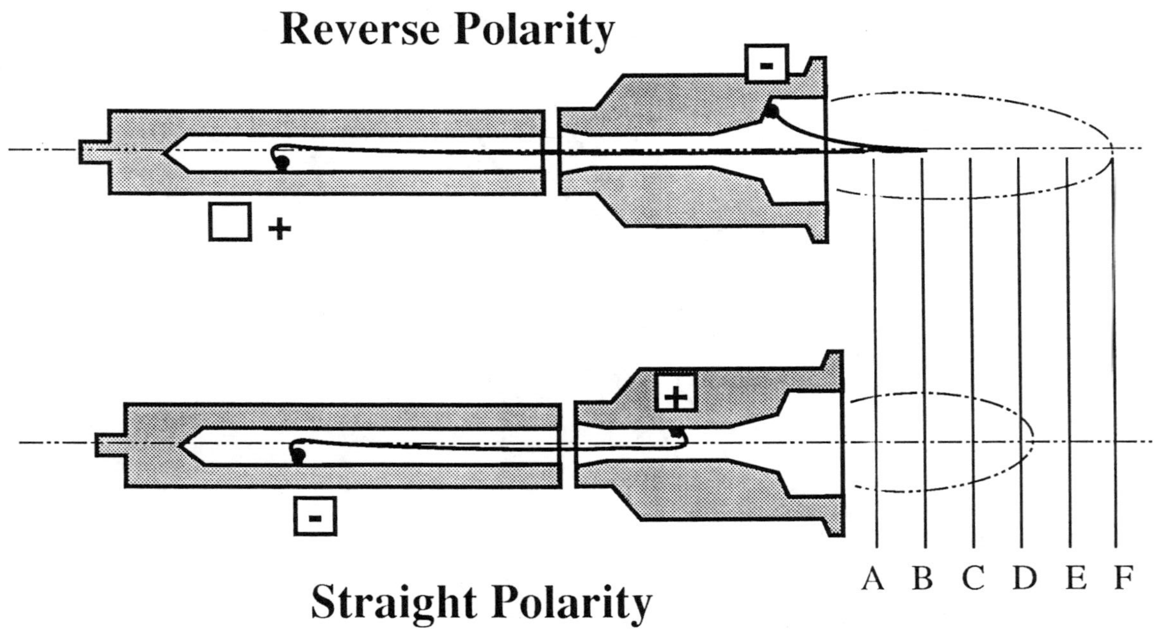

Reverse Polarity

Straight Polarity

A B C D E F

Figure 2.

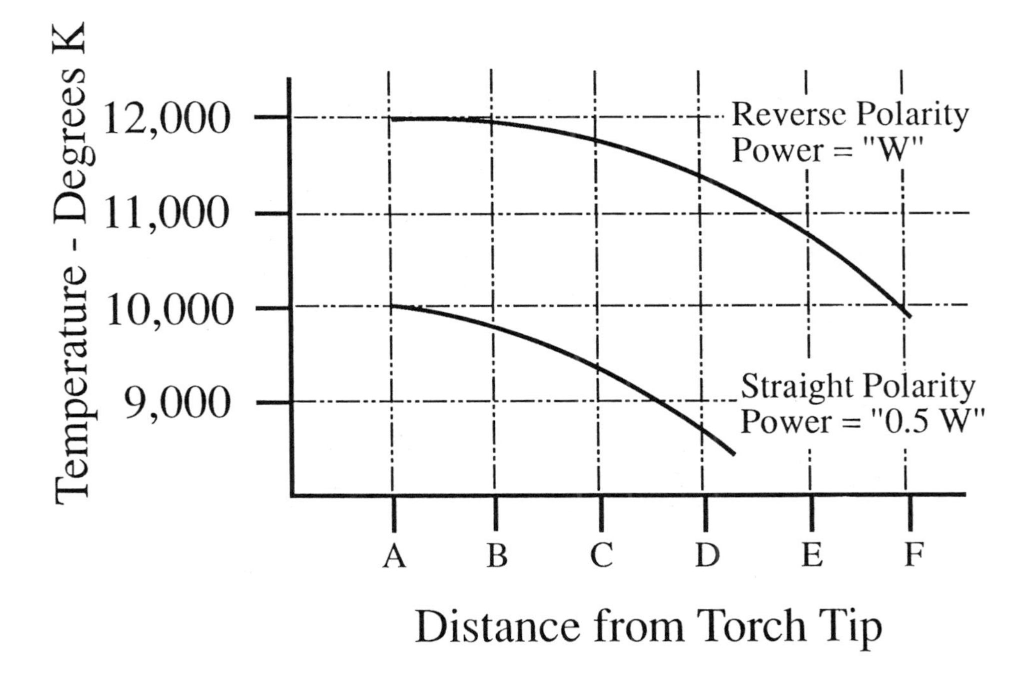
Temperature - Degrees K
12,000
11,000
10,000
9,000
Reverse Polarity
Power = "W"
Straight Polarity
Power = "0.5 W"
A
B
C
D
E
F
Distance from Torch Tip

Figure 3.

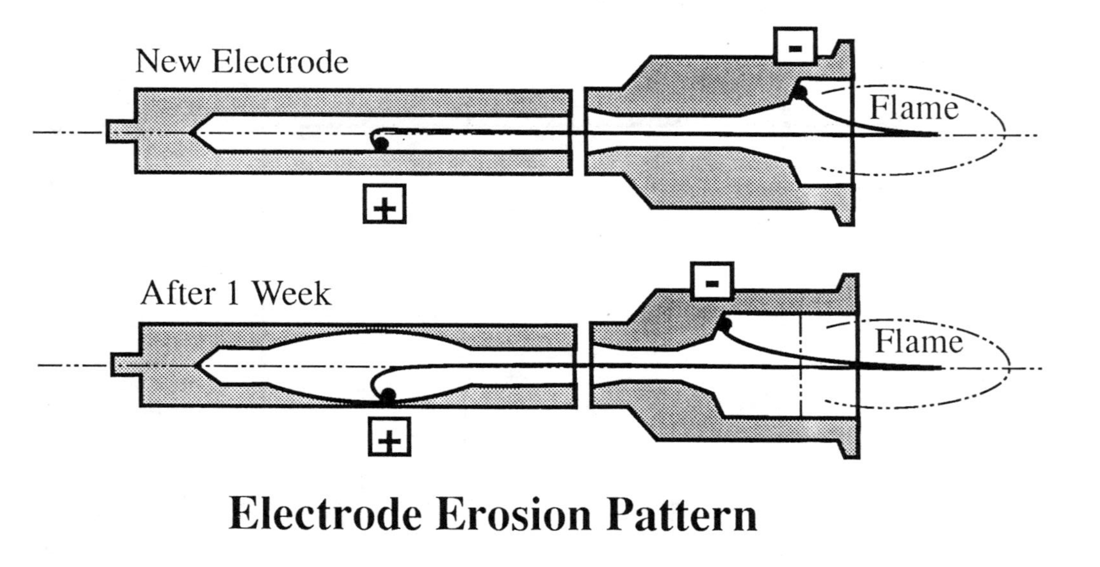

Electrode Erosion Pattern

Figure 4.

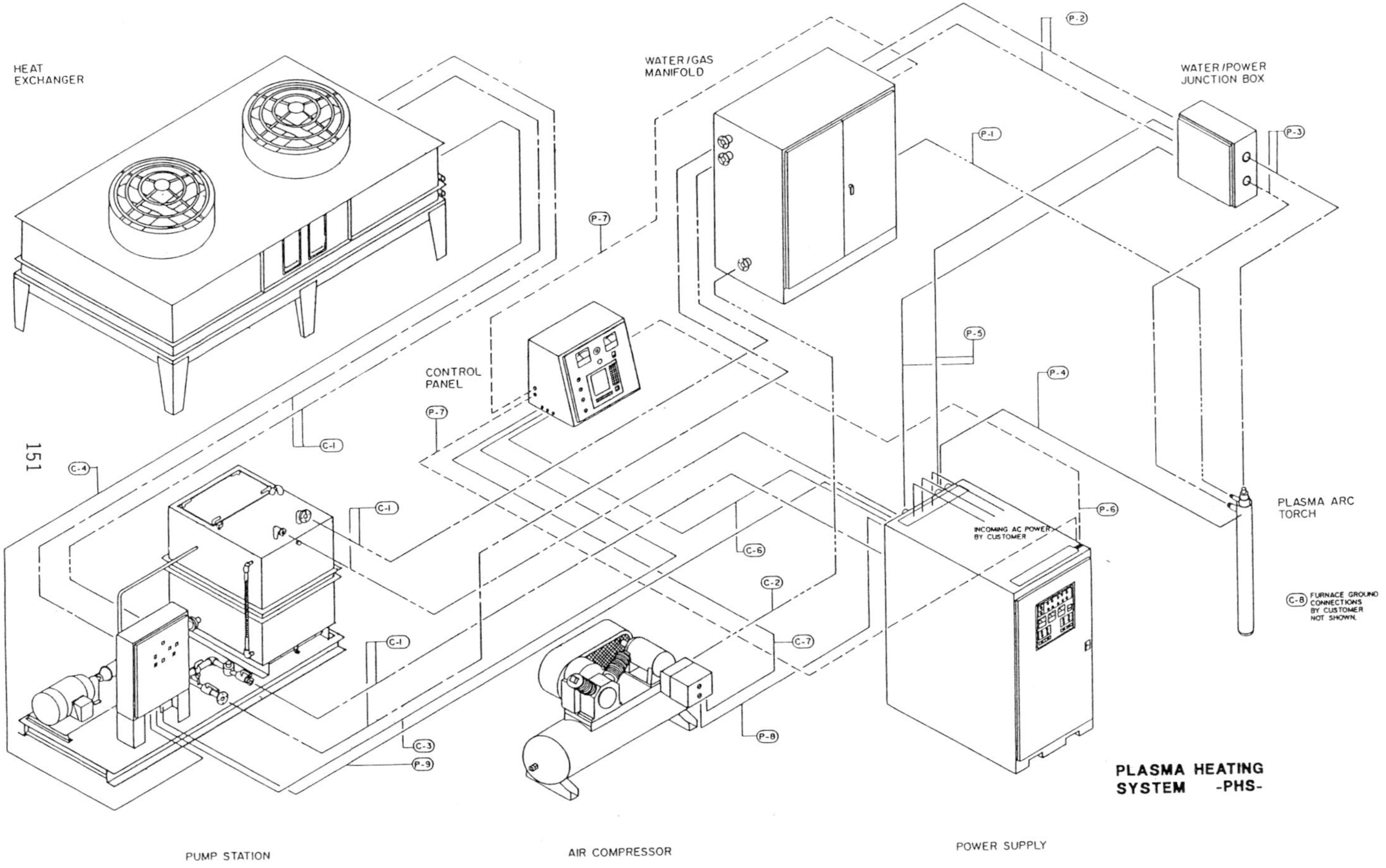

HEAT
EXCHANGER
WATER/GAS
MANIFOLD
WATER/POWER
JUNCTION BOX
CONTROL
PANEL
PLASMA ARC
TORCH
INCOMING AC POWER
BY CUSTOMER
C-8 FURNACE GROUND
CONNECTIONS
BY CUSTOMER
NOT SHOWN.
P-1
P-2
P-3
P-4
P-5
P-6
P-7
P-7
P-8
P-9
C-1
C-1
C-1
C-2
C-3
C-4
C-6
C-7
PLASMA HEATING
SYSTEM -PHS-
PUMP STATION
AIR COMPRESSOR
POWER SUPPLY

Figure 5.

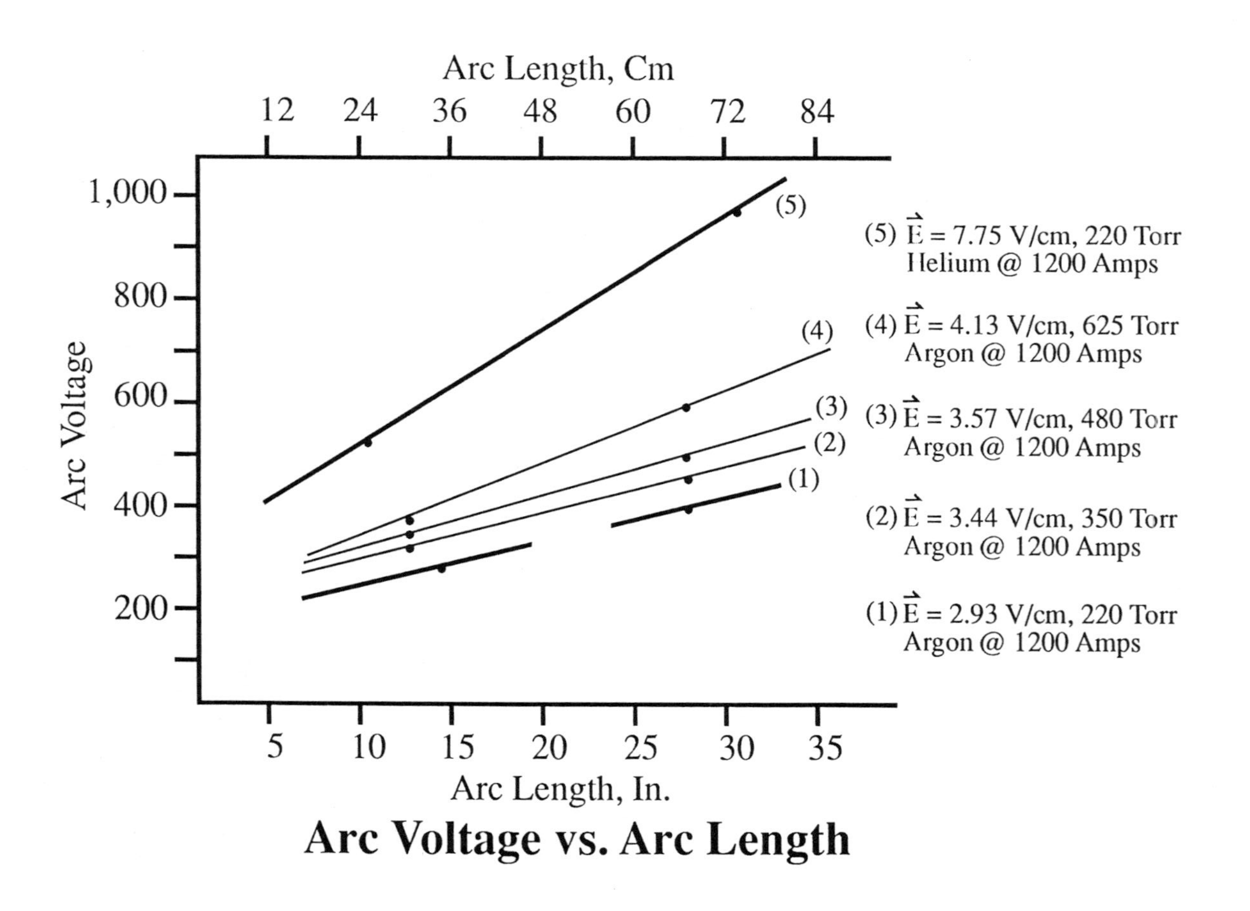

Arc Voltage vs. Arc Length

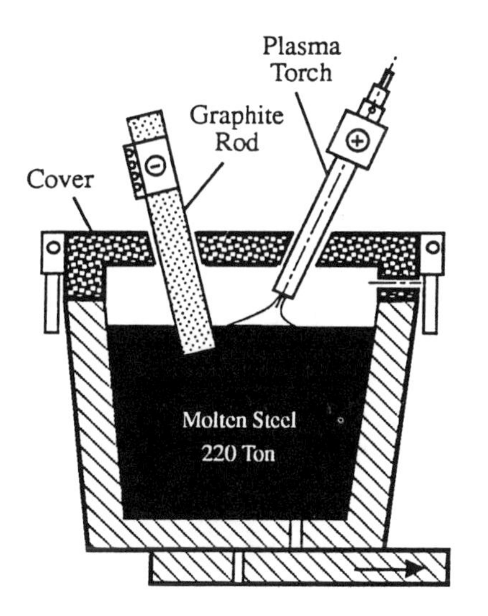

Figure 7.
USX Ladle Heater
Lorrain Works, Ohio

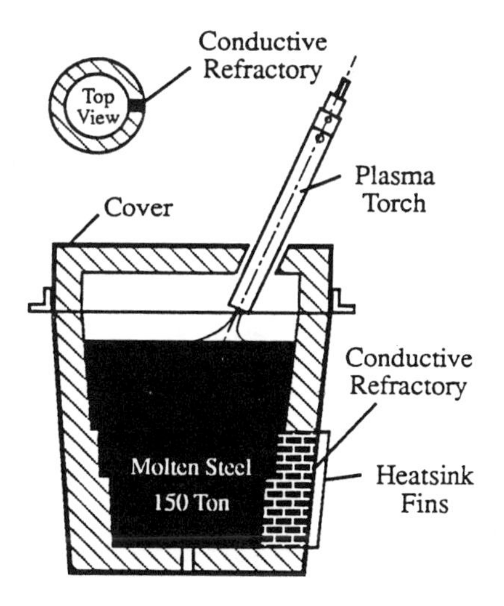

Figure 8.
Chaparral Steel Ladle Heater
Midlothian, Texas

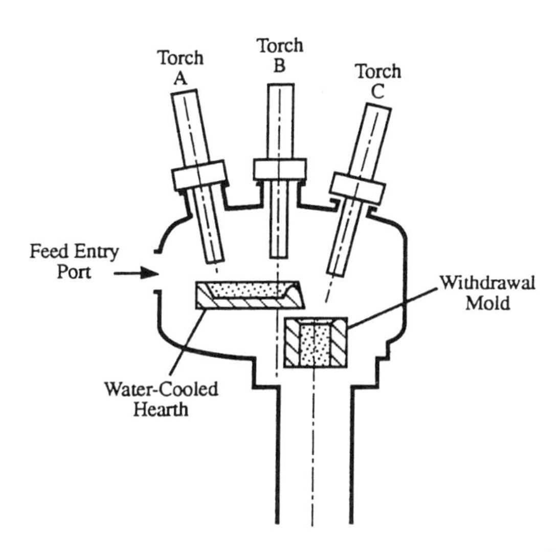

Figure 9. Plasma Furnace for Melting Titanium Scrap

MICROWAVE PLASMA PROCESS FOR THE ACCELERATED

SYNTHESIS OF NANO-STRUCTURED CARBIDES

Johanna B. Salsman*

U.S. Bureau of Mines
Tuscaloosa Research Center
P.O. Box L
Tuscaloosa, Alabama 35486-9777

Abstract

The science base has been developed to produce bulk quantities of high temperature carbides using a microwave induced plasma, MIP. Given the same throughput, high purity carbides such as WC, SiC, TiC, and TaC have been produced in an atmospheric pressure thermal CO plasma resulting in reduced processing times of approximately 99 percent requiring only 13 percent of the energy when compared to conventional processing methods. Due to the reduced processing times, the carbide powders produced in the MIP consistently displayed very uniform sizes along with nanometer microstructures.

The process was accomplished using a 915 MHz, 30 kW microwave generator operating anywhere from 5 to 7 kW resulting in plasma temperatures on the order of approximately 2,200° C. After the basics of this new technology were perfected, efforts focused on developing the technique and apparatus for industrial sized throughputs and applications. This paper describes the initial investigation into this novel carbide processing technique along with discussion on upgrading the process in both equipment and procedures to develop an industrial scale model.

* Now with U. S. Bureau of Mines, Salt Lake Research Center, 729 Arapeen Drive, Salt Lake City, Utah 84108-1283

Introduction

Tungsten carbide (WC) products continue to be the major constituent for cutting and wear resistant materials used in industrial applications. Several thousand tons of WC are produced yearly in the United States. The U.S., the largest consumer of WC powder worldwide, imported 523 tons of WC powder in 1987 and 612 tons in 1988. There are a variety of ways to produce WC but all require significant amounts of energy for extensive periods of time. Most commercial processing techniques involve heating stoichiometrically mixed amounts of elemental tungsten (W) and carbon (C) powders in electric furnaces to achieve temperatures of 1,500 to 1,600° C. The mixed powders are held at these temperatures for up to 7 hours yielding 99.9 percent pure WC. For the product to be commercially acceptable, no more than 0.06 percent free carbon is allowed.

The use of microwave energy has been examined as an alternative energy source in several processes. Some of the processes include heating of ores and materials, short pulse dielectric heating, and high temperature processing of materials. Part of this research included experiments on the synthesis of high temperature carbides since carbon is an excellent absorber of microwave energy. The goals were to reduce processing times thus reducing overall costs. The carbide compounds investigated were WC, SiC, TiC, and TaC. Techniques were developed to successfully sustain a thermal plasma in a microwave field, which led to an accelerated process to produce these carbides.

Unlike most microwave induced plasmas, this process is designed to operate at 915 MHz; allowing greater power availability to ionize a gas at atmospheric pressure. The high power ratings of the 915 MHz generators and the ability to sustain an MIP at atmospheric pressures yields higher plasma temperatures and can be considered thermal plasmas. Microwave plasmas also bear a distinct advantage over arc plasmas in that no electrodes are necessary. This one aspect of this novel process eliminates electrode contamination in the processed materials which can become a severe problem in many cases.

Equipment

Most thermal plasmas or equipment to sustain a thermal plasma generally involve an arc reactor apparatus where two high temperature electrodes are used. For this process, the thermal plasma was created using carbon monoxide gas at atmospheric pressure and a 915 MHz, 30 kW variable power microwave generator as the power source. The plasma was sustained in a chamber which was placed inside a WR 975 rectangular waveguide, a matching guide for this frequency generator. The physical dimensions of the waveguide (24.765 cm by 12.37 cm) allowed the experiments to be performed directly inside the waveguide thus simplifying the determination of the location of electric field maximums. The waveguide was short circuited at its end which produces a null of the electric field. When shorting the waveguide in this manner, the TE_{10} mode of propagation is established. Positions of electric field maximums and minimums were readily determined by measuring quarter wavelengths from the shorted end of the waveguide. The plasma chamber was placed 3/4 of a guide wavelength, λ_g, from the short in a position of an electric field maximum as depicted in figure 1.

Because the microwave was constructed of aluminum, which melts at approximately 600° C [1], it was reinforced with a graphite plate on the inside and water cooling jackets on the outside in the area where the plasma chamber was placed. Graphite was found to be a good electrical conductor therefore there were no significant changes of the electrical field patterns inside the waveguide. The plasma chamber consisted of a cylindrical zirconia crucible 7.62 cm wide and 8 cm deep with a 0.63 cm wall thickness. The crucible was placed upside down over the sample and on top of the graphite plate, as shown in figure 1. A small hole was drilled in the side of the crucible to allow the desired gas flow. The purpose of the chamber was to maintain the position of the plasma inside the waveguide and ensure that the plasma would completely envelope the sample. The natural tendency of the plasma was to stretch itself to the top of the waveguide and travel down the waveguide in the direction of the generator. Zirconia was chosen for its high melting temperature, $T_m \simeq 2,700°$ C [2], and its transparency to microwaves, $\varepsilon_r' = 7$ and $\tan\delta = 0.002$ at 915 MHz.

Two different atmospheres were maintained inside the waveguide. A positive pressure of carbon monoxide was maintained inside the chamber and was ionized by the electric field creating the plasma. Outside the plasma chamber, but inside the waveguide, a positive of helium was maintained to prevent arcing.

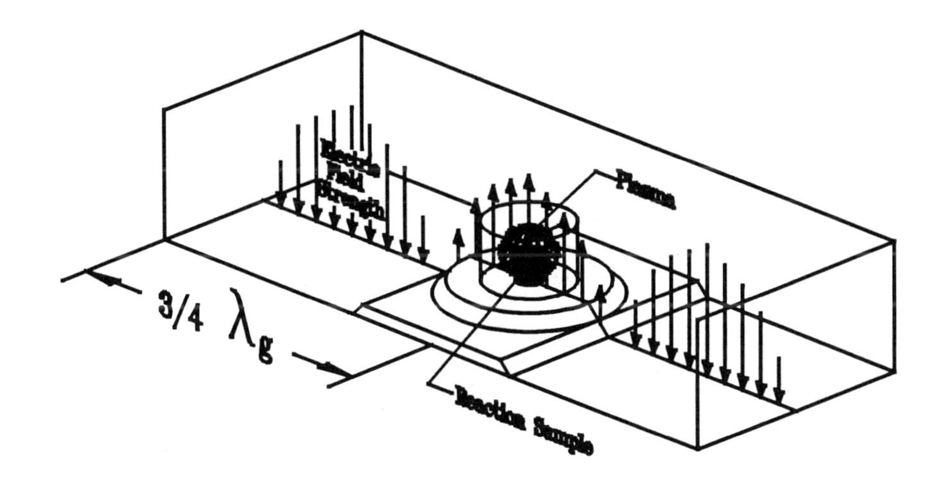

Figure 1. A three dimensional view of the waveguide showing the electric field, the crucible, the sample, and the plasma.

Procedure

For the simplicity purposes, preparation procedures and results will be given for the tungsten carbide powders only however very similar results were seen when processing the other powders mentioned above. Stoichiometric amounts of C and W powders were measured and well mixed. Approximately 50 g of the mixture was hydraulically pressed at 70 MPa, forming a 2.54 cm diameter pellet 2 cm in height. The pellet was then encapsulated in carbon and repressed. The use of excess carbon, which must be subsequently removed to ensure a commercially acceptable product, would not be necessary if trace amounts of air were not present in the waveguide. However, the experiments were conducted under conditions such that it was difficult to ensure that oxygen was not present even after flushing the waveguide with helium. Because of the extreme temperatures created by the plasma, the presence of oxygen tended to oxidize carbon from the stoichiometrically mixed sample, thus causing an incomplete reaction. After the experiment was complete, the excess carbon flaked off the WC pellet quite easily.

Once the sample was prepared, it was placed inside the waveguide and in the plasma chamber at the position of a previously determined electric field maximum. The positioning of the chamber is illustrated in figure 1. The waveguide was closed and sealed by placing the shorting plate on the end. Next, the waveguide and plasma chamber were flushed with a positive pressure of helium and carbon monoxide respectively, for approximately 20 to 30 minutes which removed most of the oxygen present. After flushing the waveguide for the prescribed time, 5 to 10 kW of microwave power was applied to initiate the plasma. The plasma ignited instantaneously. The plasma was sustained around the sample for 10 minutes at 7 kW of power and then extinguished by terminating the power. The sample was allowed to cool for 30 minutes in the waveguide filled with a helium atmosphere to prevent oxidation during cooling. Approximately 50 g of commercially acceptable WC was produced.

Results

X-ray analysis of the WC product indicated only WC present, with no di-tungsten carbide, W_2C, or elemental W detected (W_2C and W in the final product indicate an incomplete reaction). X-ray analysis did not detect carbon, and chemical analysis to detect trace amounts of free carbon were performed on the samples. Results of these tests showed only a 0.04 percent free C content indicating a product of commercial quality. These experiments were repeated resulting in similar chemical analyses.

Particle size analyses were performed on WC powders produced in the MIP and on samples of commercially produced WC for comparison. The MIP produced WC particle sizes ranging from 0.5 to 4 μm with 50 percent of the particles less than 2 μm. The commercially produced WC particles varied in size from 1 to 10 μm with 50 percent less than 5 μm. Examination by scanning electron microscopy revealed large agglomerates of various size particles in the commercially produced WC. These agglomerates were composed of particles that had a maximum cross section of 17 μm and were loosely spaced from each other. Conversely, the agglomerates of the WC produced in the MIP were tightly packed clusters of extremely fine particles, from 0.5 to 3 μm.

A cobalt (Co) binder was added to the WC powder produced in the MIP to form a 10 percent Co - 90 percent WC mixture. The WC-Co mixture was isostatically pressed into a 1.25 cm diameter pellet. The pellets were sintered at 1,350° C for 1 hour. After sintering, the pellets were polished and then etched with Murikami's Reagent [3]. The etching conditions ranged from 2 to 20 minute exposure to the etchant at 20 to 50° C. The etched pellets of WC were dried over night and then examined with the SEM. A photomicrograph of the WC after etching revealed grain sizes on the order of 0.05 μm as is shown in figure 2. Literature sources [4] have reported sintered WC grain sizes on the order of 0.2 to 0.3 μm.

It has frequently been reported for commercial production of ceramics that shorter processing times reduce grain growth and segregation of impurities to the grain boundaries [5]. It is believed that the smaller particles found in the WC powder produced in the MIP is a result of the much shorter processing times required by the MIP process.

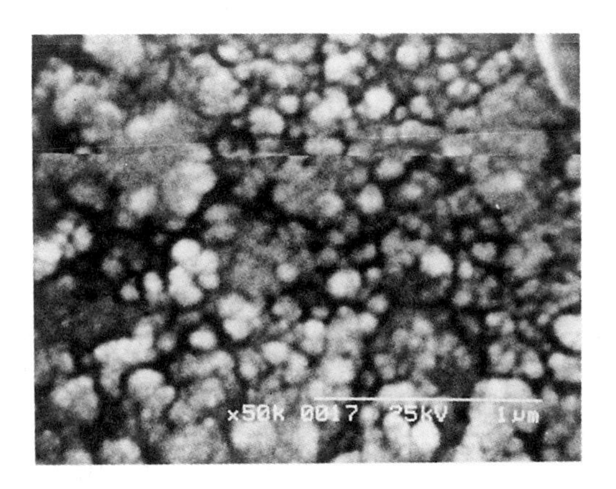

Figure 2. SEM photomicrograph of etched WC magnified 50,000 times. Bar denotes 1 μm.

The energy costs for the MIP process were compared to conventional WC processing techniques. The assumptions made in this comparison are as follows: (1) The heat loss from both methods is by convection and is proportional to the difference in the furnace and ambient temperature. (2) The heat losses from either system are the major energy use, and they are large compared to the amount of heat required to raise the sample temperature; thus the latter can be neglected; (3) The equipment parameters are the same.

From these calculations, it was found that the energy costs of producing WC by conventional methods is approximately \$8.50/kg compared to \$1.10/kg for production of WC in the MIP. Furthermore, the reduction in processing time by a factor of nearly 40 indicates that the microwave system needs only 2 to 3 percent the capacity of the furnaces used in conventional methods in order to achieve the same throughput.

Pilot Scale Model

It was shown that the use of a MIP to produce nano-structured high temperature carbide powders was feasible but only on a small scale. The next step in the experimental procedure was to design and construct a large volume MIP cavity where 5 to 10 kg of material could be produced at one time. Only an apparatus such as this could make this novel processing technique acceptable to the powder processing manufacturers. It is one thing to develop a new processing method and yet another to enlarge the process for manufacturing purposes. The principles of the large volume cavity are the same as for the waveguide technique, but the dimensions and volume had to be increased. It became no longer feasible to operate the plasma in the waveguide therefore a plasma cavity was designed and constructed. Specifications of the cavity were quite arduous in that it needed to withstand the ultrahigh temperatures generated by the plasma, allow an inlet and outlet to place the materials inside and out of the cavity, and at the same time be sealed for microwaves and gas. Such a system was designed and is shown in figure 3.

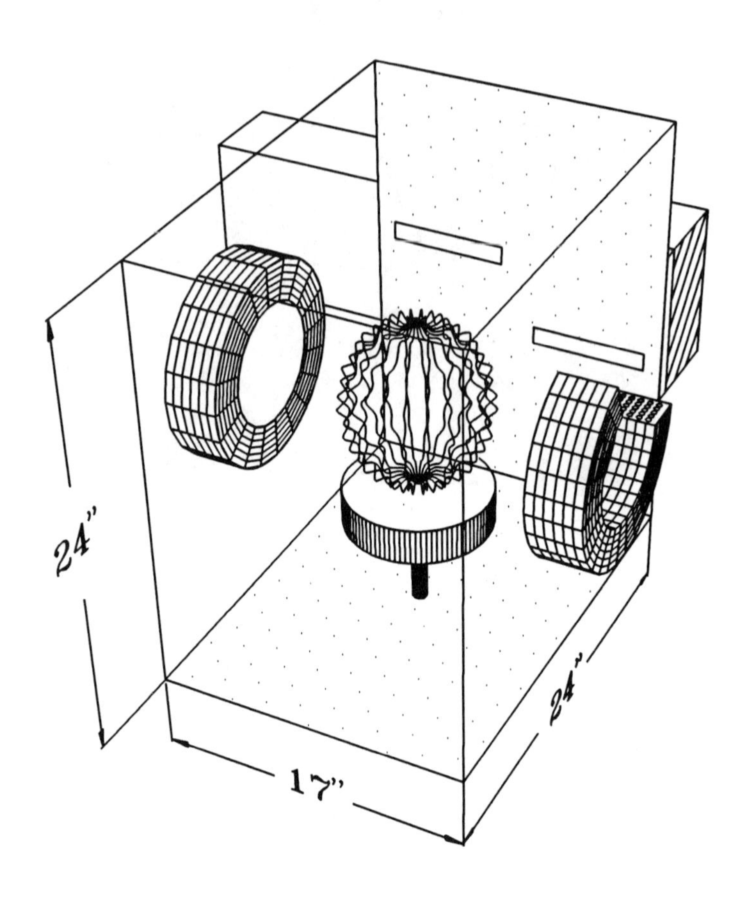

Figure 3. A three dimensional drawing of the high temperature large volume plasma cavity.

The DC magnets shown on each side of the cavity are used to shape the electric plasma over the sample. A large portal was designed into the cavity to allow specimens to be placed into or withdrawn from the cavity and also with the fused quartz window allowed viewing during plasma operation. A sample plate made of graphite was positioned in the area where the plasma would most likely occur and could be raised or lowered at whatever position necessary for maximum plasma exposure. Experiments with material processing inside the large volume cavity have yet to be performed however work is continuing in this area. The cavity itself is capable of sustaining a large volume plasma and performs as was designed. It is the belief of this researcher, that once the cavity has been thoroughly tested, large volumes of nano-structured carbides can be produced inside a microwave cavity such as this which shall save both time and energy in the overall production of these materials. Figure 4 is a photograph of the plasma cavity showing a CO plasma.

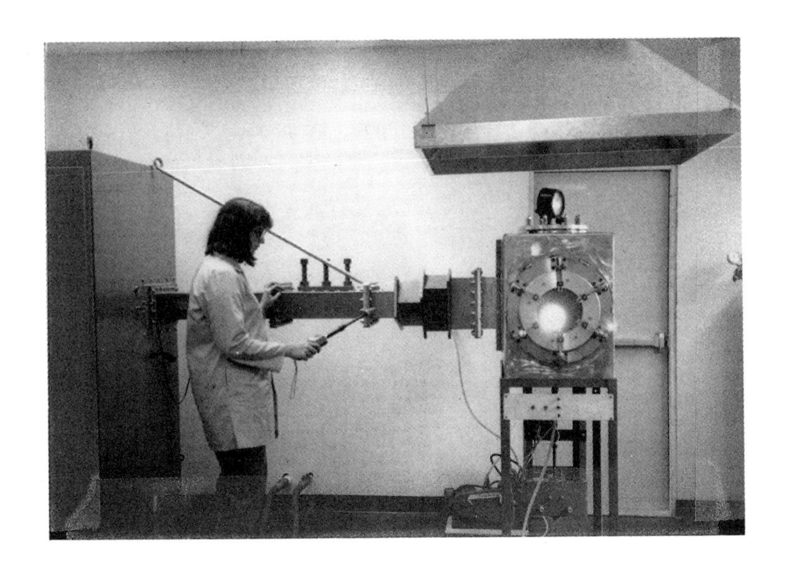

Figure 4. A photograph of the plasma cavity.

Conclusions

Microwave plasmas are successfully being used for depositing thin to thick films, such as diamond coatings, on various materials [6]. However these plasmas have not been used for mass producing high temperature materials until now. The science base for a new process to produce WC in less than 3 percent of the production time required by conventional methods and 13 percent the energy has been developed. The key element of this technology process utilizes the heat generated from a MIP that is capable of temperatures in excess of 3,000° C. Indications are that a shorter processing time, or less time at temperature, reduces grain growth, thus yielding a powder that is finer grained. These finer precursor powders should improve the overall strength and toughness of the finished carbide product.

References

1. Robert C. Weast, ed.-in-chief, Melvin J. Astle, and William H. Beyer, eds. CRC Handbook of Chemistry and Physics, 69th Edition, (CRC Press, Inc., Boca Raton, FL. 1988) p. B-7.

2. R. C. Weaste, ibid., p. B-145.

3. E. M. Ekmajian, and J. B. Bulko "Preparation Techniques for Structural Characterization of Powdered and Composite Materials," (Mat. Res. Soc. Symp. Proc. 115, Reno, NV 1988) pp. 87-92.

4. Upadhya, K. "An Innovative Technique for Plasma Processing of Ceramics and Composite Materials." Am. Ceram. Soc. Bull., v. 67, No. 10, 1988, pp. 1691-1694.

5. G. Petzow, Metallographic Etching. (Pub. by ASM, Metals Park, Ohio, 1978) p. 65.

6. Hopwood, J., Reinhard, D., and Asmussen, J. "Performance of Multipolor Electron Cyclotron Resonant Microwave Cavity Plasma Source." Paper in the Proc. of the 24th Microwave Power Symposium (Stamford, CN, Aug. 28-30, 1989). Int. Microwave Power Institute, Clifton, VA, 1989, p. 27.

Section IV

A THERMODYNAMIC ANALYSIS OF TITANIUM CARBIDE SYNTHESIS IN A THERMAL PLASMA REACTOR.

R.L. Stephens, M.K. Wu and B.J. Welch
Department of Chemical and Materials Engineering

J.S. McFeaters
Department of Mechanical Engineering
The University of Auckland
Private Bag 92019
Auckland
New Zealand

J.J. Moore
Department of Metallurgical and Materials Engineering
Colorado School of Mines
Golden CO 80401
U.S.A.

Abstract

The synthesis of TiC in a thermal plasma reactor was studied using free energy minimisation to calculate equilibrium states of the system over a range of temperatures, pressures and compositions. Reactants considered are TiO_2 and CH_4 in an Ar plasma. Consideration was made of kinetic factors including nucleation, residence time and reaction rates. The inclusion of the inert gas in the system specification was shown to have a significant influence on the system, which is similar to lowering the overall system pressure. However, this effect does not correspond to simply lowering the partial pressure of the reactants. Therefore inert gases must be included explicitly in the system specification. Results are presented in species versus temperature plots and condensed-phase stability diagrams. TiC synthesis from TiO_2 and CH_4 in a thermal plasma torch is shown to be thermodynamically feasible within the limits of the process variables considered.

Introduction

The production of ultrafine ceramic powder in thermal plasma reactors has been the subject of research for many years [1-3], and examples of commercial-scale processes can now be found [4]. Thermal plasma reactors most commonly use either a radio-frequency (RF) electromagnetic field or an electric arc struck between two or more electrodes to generate temperatures in the order of 8000 K to 15000 K. Residence times are typically on the order of 1-10 ms. At these temperatures, with a well designed system, all the reactants are vaporized and the system is comprised almost entirely of ions, atoms, and electrons. Product formation takes place after the mixture leaves the high temperature zone and begins to cool. During this period, temperatures are still high enough (on the order of 2000 to 5000 K) so that most chemical reaction rates are collision limited rather than energy limited. The resulting high reaction rates serve to maintain the system at or near gas phase chemical equilibrium until the system temperature falls to the order of 1500 to 3000 K. Thus equilibrium modelling can be used to follow the progress of gas-phase chemistry down to temperatures below where the first condensed phases have appeared.

Ceramic products have relatively low vapour pressures, and as the gas mixture cools, it is possible to generate very highly supersaturated mixtures leading to the formation of ultrafine particles by homogeneous nucleation. When the nucleation step is followed by a rapid quenching of the flow, the desired products can essentially be frozen in so that the high temperature condensed phase remains as the system temperature decreases, rather than being consumed by condensed-phase reactions as predicted by equilibrium thermodynamics. Significant deviations from equilibrium occur for condensed phases because gas-solid and gas-liquid reactions generally proceed at lower rates than gas-phase reactions, as they are usually limited by the diffusion of reactants and products through the particle and the boundary layer surrounding the particle. Further, at very high quenching rates, the gas phase chemistry may also deviate from equilibrium. Thus, the chemistry of the system and particle size distributions can be manipulated to some extent by controlling the quenching rates of cooling flows using either gas dynamic techniques [5] or by the injection of cold quench gases [6].

It is useful to have predictive tools available for narrowing the range of variables to be investigated when attempting to isolate the optimum reaction stoichiometry for a reaction system in a thermal plasma reactor. Two approaches may be used to attempt to predict the products of a reaction system. They are equilibrium modelling and kinetic modelling. Equilibrium modelling of thermal plasma reactors for the production of ultrafine ceramic powders has been performed by many investigators. The most common technique for equilibrium modelling is free energy minimisation of a specified reaction system. Examples of systems that have been examined using this technique include carbothermal reduction of silicon dioxide [7] and aluminium oxide [8], and the synthesis of silicon nitride [9] and silicon carbide [10]. The attraction of this technique is its simplicity. Equilibrium of multi-component, multi-phase systems can be calculated using a number of easily-obtainable computer programs such as the CSIRO-THERMODATA suite of programs [11], and other similar packages [12].

An important factor which has often been overlooked in plasma thermochemistry calculations is the effect of inert gases. In thermal plasma synthesis the gases are comprised largely of inert gases which form the plasma and convey reactants. The reactants are typically only a small proportion of the total gas content (around 0.1 mol% to 1 mol%), especially in laboratory-scale reactors and the partial pressures of reactants and products are correspondingly low.

Inert gas addition affects the extent of reactions in a way similar to Le Chatalier's principle. When the net change in the number of moles going from reactants to products increases, the product yield is increased with decreasing pressure. Inert gas addition reduces the relative net increase in mole number which will have a similar effect to reducing the total system pressure. Thus, when the net change in the number of moles is positive, inert gas dilution will increase the yield.

Titanium Carbide Production

TiC_x is commercially produced by solid-phase reaction between intimate mixtures of TiO_2 and carbon black at 2000 °C or more under a H_2 atmosphere. Storms [13] discusses this process in terms of the typical process conditions, and their effect on the product, and also, other less common processes.

Thermal plasma processing is an alternative route for TiC_x production. To date, relatively little work has been performed on DC or RF plasma synthesis of TiC_x. Ishizaki et al. [14] produced 50 to 100 nm cubic phase TiC_x particles in an arc plasma using Ti and CH_4 as reactants. Neuenschwander [15] prepared ultrafine TiC_x powders in the size range 10 to 100 nm by reaction of $TiCl_4$ with CH_4 in a H_2 DC plasma jet. The powders were slightly sub-stoichiometric with a carbon content of 18.0 ± 0.5 mass% compared with a theoretical carbon content of 20.05 mass% for $TiC_{1.0}$. This may be due to partial reduction of the TiC_x by the hydrogen atmosphere or by the removal of reactive carbon donating species by reaction of these species with the hydrogen atmosphere. The powders contained 1.4 ± 0.2 mass% O contamination and x-ray diffraction yielded a composition for the powder of $TiC_{0.90}O_{0.05}$.

Mitrofanov et al. [16] used a DC arc plasma reactor to react CH_4 with both Ti and TiO_2 powders to synthesise TiC_x. The TiC_x produced from Ti metal powder was found to have lattice parameter of 4.33 Å, corresponding to a composition of $TiC_{0.93}$ [17], although the oxygen level was not determined and this may have affected the measured lattice parameter [18]. The TiC_x produced from TiO_2 was very similar to those observed with synthesis from Ti metal powder. The carbide powders were very fine (100 Å) with a lattice parameter of 4.328 Å.

Method of Analysis

Calculation of the equilibrium composition of a specified system were performed using the CHEMIX program of the CSIRO-THERMODATA [11] suite of computer programmes. The programme uses the technique of minimisation of Gibbs free energy to determine the equilibrium composition of the system at the specified temperature and pressure, subject to materials balance constraints. The advantage of this method is that it does not require a knowledge of the reaction mechanisms for the system; only the set of participating species (reactants and possible products) and their thermodynamic properties need be known.

Thermodynamic property data were obtained from the JANAF Thermochemical Tables [19]. All relevant species included in the JANAF Thermochemical Tables were included in the species lists for the input files. Data for $TiC_{(g)}$, which are listed in Table I, were estimated using statistical thermodynamics and estimates of spectroscopic data and molecular orbital calculations (McFeaters [20], Herzberg [21] and Bauschlicher [22]). There was no thermodynamic data available for the solid solution TiC_xO_z; thus this condensed phase was not included in the calculations. No allowance was made for any deviations of $TiC_{(s)}$ from the

Table I: Spectroscopic Constants of TiC.

Vibrational constant $\tilde{\omega}$ (cm^{-1})	950
Rotational constant B (cm^{-1})	0.56
Interatomic Spacing r_e (Å)	1.77
Dissociation energy ΔD_e (kJ Mol^{-1})	530
Electronic ground state degeneracy g_{e0}	4

stoichiometric composition of $TiC_{1.0}$ and the only condensed-phase allotrope of carbon considered was graphite.

The programme was run for temperatures between 1000 and 6000 K, in 50 K intervals, and at a pressure of 1 atm unless otherwise specified. THERMODATA output files were post-processed and data extracted only for those species with number of moles greater than 0.001.

Analysis of the Plasma Synthesis of Titanium Carbide

Table II: Ar/TiO_2 molar ratios and the corresponding TiO_2 feed rate for a RF plasma torch running on 25 l min^{-1} Ar.

Ar/TiO_2 Molar Ratio	TiO_2 Feed Rate (g min^{-1})
1	89.2
10	8.92
100	0.892
1000	0.089

An important aspect of thermochemical modelling of plasma-based ceramic production systems which has often been overlooked is the effect of plasma gases on the system equilibrium. Plasma gases are usually inert and are present only to sustain the plasma. In the case of a 15 kW laboratory-scale RF plasma torch, a typical total argon flow rate would be 25 l s^{-1} (STP). Table II lists TiO_2 feed rates corresponding to a logarithmic progression of argon to TiO_2 molar ratios. Typical powder feed rates used in plasma synthesis of ceramic powders range between 0.1 g min^{-1} and 5 g min^{-1} for a 15 kW RF plasma torch. These feed rates correspond to argon to TiO_2 molar ratios of 892 to 18 respectively.

The system under consideration in this paper is the Ar-Ti-C-O system, in which titanium dioxide (TiO_2) is carbothermally reduced with methane (CH_4) in an argon (Ar) plasma to synthesise titanium carbide (TiC_x). The overall stoichiometric reaction equation for this process is:

$$TiO_2 + 3CH_4 \rightleftharpoons TiC + 2CO + 6H_2 \quad (1)$$

Stability diagrams have been calculated for titanium carbide synthesis from TiO_2 and CH_4 at argon to TiO_2 molar ratios of 1, 10, 100 and 1000. The results of these calculations are presented in Figures 1 to 4.

Analyses of chemical vapour deposition systems for the production of Si_3N_4 [23] and SiC [24] led Kingon et al. to conclude that higher concentrations of inert gases have an effect similar to slightly lowering the system pressure. However a system which has been diluted with an inert gas differs from one in which the system pressure has been lowered in that the rate at which equilibrium is approached will be different. This is due to the effects of different gas collision rates and mean-free paths in the gaseous phase on the transport rates. This point is especially relevant for plasma reactors where the system may deviate from equilibrium in regions where the thermochemistry is of interest. Another important aspect of inert gas dilution is the effect on homogeneous nucleation of species with high latent heats. Dilution from the inert gas will reduce the temperature rise of the nascent clusters as they form from the gas phase and also mollify any bulk temperature changes caused by condensed phase transformations. This can greatly enhance particle nucleation and growth rates by keeping condensed phase temperatures down and increasing condensation rates.

To compare the effect of lowered system pressure to the effect of inert gas addition, stability diagrams were prepared for system pressures of 0.1, 0.01 and 0.001 atmospheres. The diagrams corresponding to these reduced system pressures are presented in Figures 5 to 7 respectively. At a system pressure of one atmosphere, there was no substantial difference between the stability diagram calculated for no argon and that for one mole of argon. Therefore Figures 5 to 7 can be compared to Figure 1.

There are several features common to each set of diagrams. Firstly, only for a stoichiometric ratio of 3 moles of CH_4 to 1 mole of TiO_2 is $TiC_{(s,l)}$ the only predicted condensed phase in the temperature window examined. For all other ratios, a second condensed phase is thermodynamically predicted to form, leading to a two condensed-phase system. A possible exception would be the formation of TiC_xO_z solid solutions but for the sake of comparison, they are discounted.

It can be argued that the formation of the oxycarbide can be discounted for molar ratios of CH_4 to TiO_2 of greater than approximately 2 because $TiC_{(s)}$ is always predicted to form well before to $TiO_{(l)}$. Therefore one could reasonably expect a two phase mixture of $TiC_{(s)}$ and $TiO_{(l)}$ to form rather than a solid solution since the time available for diffusion between the two phases would typically be short due to the high cooling rates experienced by condensing particles [25]. However, the assumption of no solid solution may not be valid for molar ratios of less than 2 since either $TiO_{(l)}$ is predicted to condense first followed by the solid phase $TiC_{(s)}$ or the two phases are predicted to condense simultaneously. In this case, a solid solution could be formed by co-condensation, the composition of which would depend on the local concentrations of the gaseous precursors.

Secondly, the temperature at which the first condensed phase is predicted to appear decreases for both increasing argon to TiO_2 molar ratios and decreasing system pressures. The decrease in the temperature at which the first condensed phase is predicted to appear has ramifications for any theoretical analysis of the condensation process and may influence the particle size distributions obtained experimentally by reducing the degree of effective sub-cooling that a process can achieve. Reducing the degree of sub-cooling will then result in the formation of fewer nuclei in the system during the condensation process and ultimately larger particles.

Thirdly, the stability of $TiC_{(l)}$, as indicated by the width of the temperature window that the phase is predicted to exist in, also decreases for both increasing argon to TiO_2 molar ratios and decreasing system pressures. The elimination of $TiC_{(l)}$ in favour of $TiC_{(s)}$ as the first predicted condensed phase will almost certainly change the morphology of the primary crystallites due to the elimination of the influence of surface tension on liquid particles causing the primary crystallites to become spherical. Following the arguments made previously for the lack of time for significant solid state diffusion, one could expect agglomerates formed from crystallites that

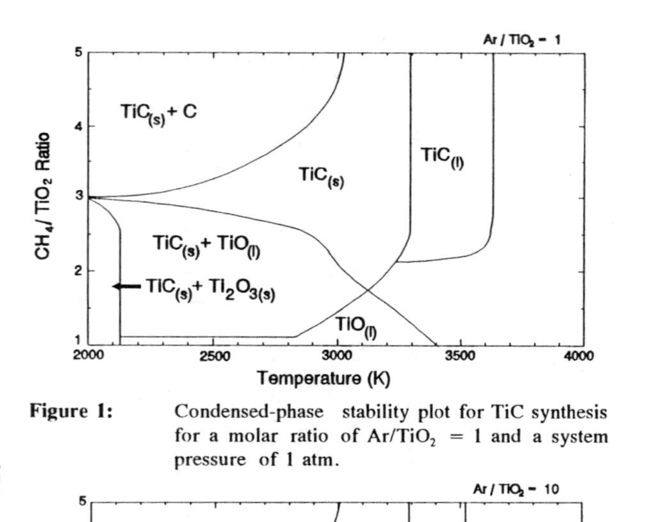

Figure 1: Condensed-phase stability plot for TiC synthesis for a molar ratio of Ar/TiO_2 = 1 and a system pressure of 1 atm.

Figure 2: Condensed-phase stability plot for TiC synthesis for a molar ratio of Ar/TiO_2 = 10 and a system pressure of 1 atm.

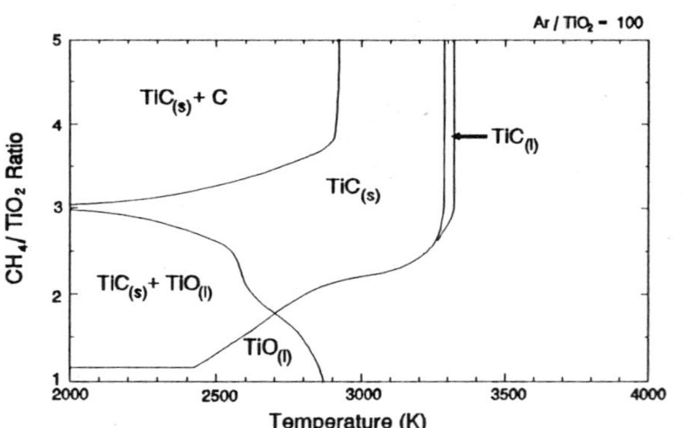

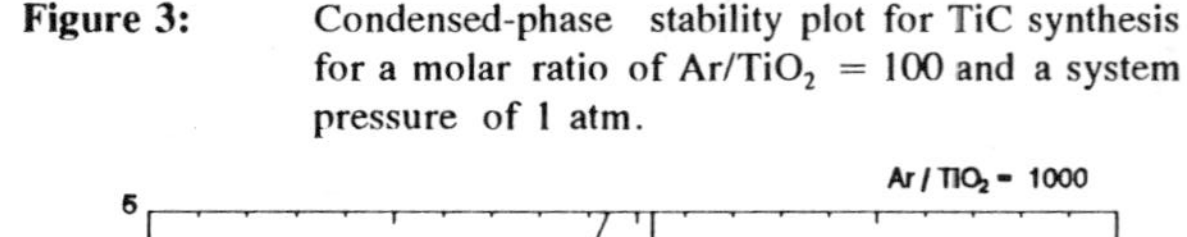

Figure 3: Condensed-phase stability plot for TiC synthesis for a molar ratio of Ar/TiO_2 = 100 and a system pressure of 1 atm.

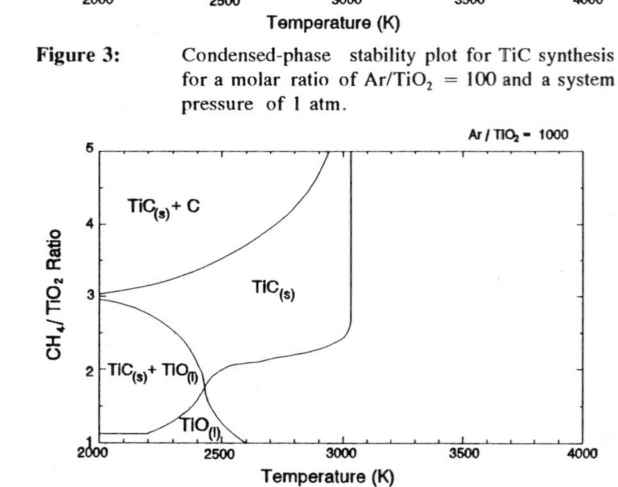

Figure 4: Condensed-phase stability plot for TiC synthesis for a molar ratio of Ar/Ti = 1000 and a system pressure of 1 atm.

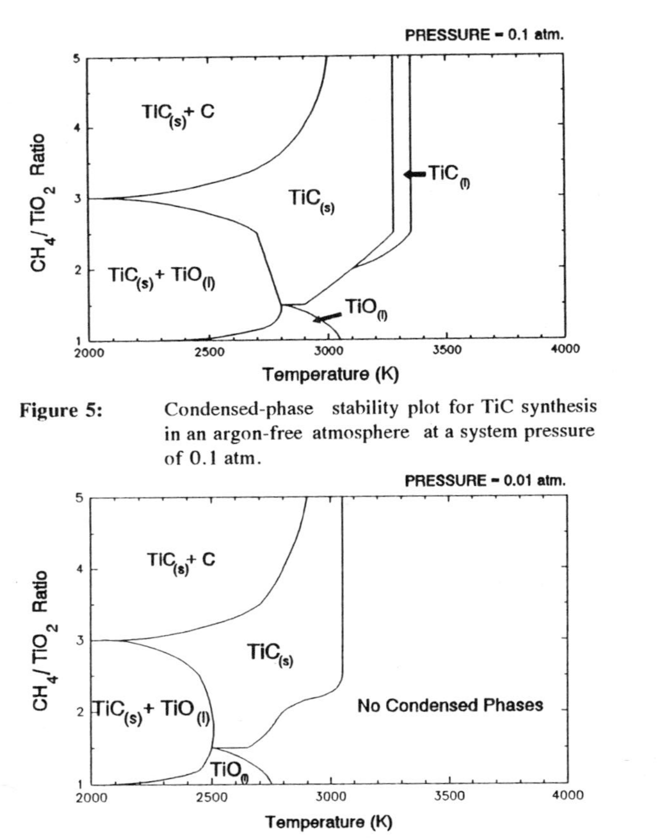

Figure 5: Condensed-phase stability plot for TiC synthesis in an argon-free atmosphere at a system pressure of 0.1 atm.

Figure 6: Condensed-phase stability plot for TiC synthesis in an argon-free atmosphere at a system pressure of 0.01 atm.

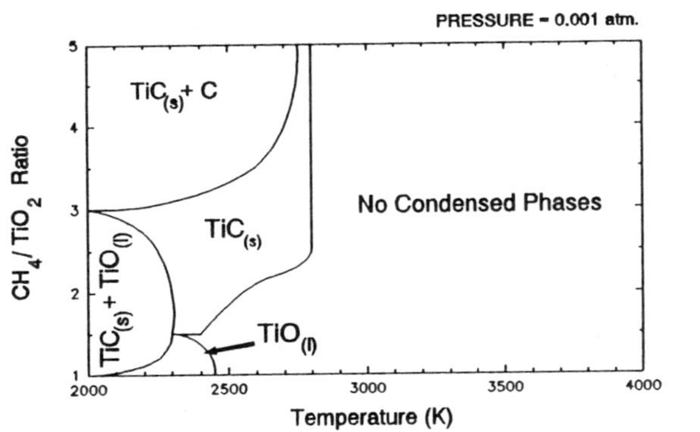
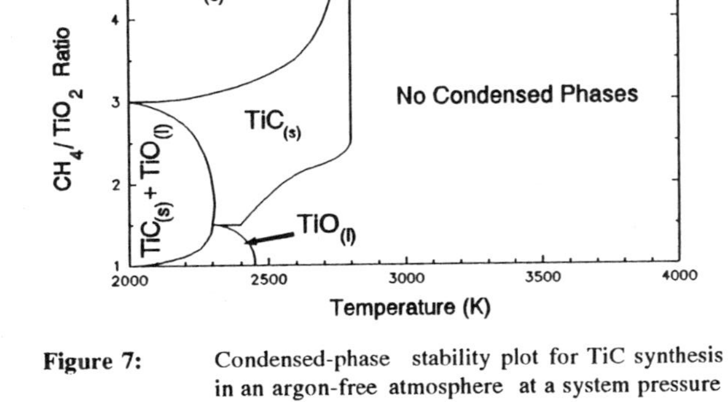

Figure 7: Condensed-phase stability plot for TiC synthesis in an argon-free atmosphere at a system pressure of 0.001 atm.

did not form via the liquid phase to be less dense since the primary crystallites are not as rounded and their microscopic protuberances would result in reduced packing densities compared to spherical particles. This would have an important influence on the quality of the agglomerates as a powder for the formation of advanced ceramic components since there would be increased dimensional change accompanying the sintering process.

At the lower end of the temperature spectrum, titanium suboxides are only observed for two diagrams. This does not indicate that these phases are not predicted to appear; merely that they are not predicted to form within the temperature range considered for the preparation of these diagrams. However with the rapid cooling of the particles in the plasma reactor causing deviations from equilibrium due to the elimination of reactions between condensed phases, these phases should not appear in the powder produced by a plasma process.

The analyses presented in this manuscript have shown that TiC synthesis from TiO_2 and CH_4 in a thermal plasma torch is thermodynamically feasible within process condition windows identified by the analyses. The effects of the various process conditions on the expected condensed phases do not vary widely over the inert gas to TiO_2 molar ratios or the system pressure. For CH_4 to TiO_2 molar ratios from just above 1.0 to approximately 1.7, the predicted products are TiO with trace amounts of TiC, provided the cooling rates are high enough to prevent condensed-phase reactions. From a molar ratio of 1.7 to 3.0, the predicted products are TiC with trace amounts of TiO while above molar ratios of 3, the predicted products are TiC and soot.

The reduction in the temperature at which the first condensed phase is predicted to form and the elimination of suboxides from the diagrams is part of a general trend of the reduction in the temperature of all phase transitions with increasing argon to reactant molar ratios or with decreasing system pressure. The only exception to this statement is the formation of solid carbon for molar ratios of methane to TiO_2 of greater than 3. The temperature at which solid carbon, in the form of graphite, is predicted to form is largely independent of the ratio of argon to reactants or system pressure. For instance the temperature at which carbon is predicted to appear for a ratio of 5 moles of methane to 1 mole of TiO_2 only decreases from 3025 K for only 1 mole of argon and a system pressure of 1 atmosphere to 2940 K for 1000 moles of argon and 1 atmosphere or to 2750 K for no argon but a system pressure of 0.001 atmospheres.

The relative lack of influence of inert gas addition or system pressure on the formation of condensed carbon has important ramifications for the synthesis of titanium carbide powder in a plasma reactor. This is because the stoichiometry in such a system is almost always carbon-rich to try to force the synthesis reactions to completion. As the width of the temperature window between $TiC_{(s)}$ and carbon decreases, it is increasingly less likely that the arguments proposed by Chang and Pfender [26] would be valid with respect to condensed carbon phases not forming particularly since soot formation may well occur through other mechanisms. Therefore the quality of the powders produced would be reduced because they could be expected to contain significant amounts of free carbon.

An important point should be made regarding the thermodynamic data used in the calculations and the influence this has on the predicted products. The only data included in the JANAF Thermochemical Tables [19] for solid carbon is for the graphite allotrope. According to the equilibrium criterion [26] used to decide whether a species will condense, graphite is not predicted to appear in the condensed-phase products. Graphite has not been experimentally observed in titanium carbide synthesis experiments performed by the authors. However, significant contamination of the condensed-phase products by amorphous carbon soot has been observed.

The mechanism for soot formation is significantly different from that for the formation of graphite from the gas phase. For graphite to form, carbon must reach a high enough supersaturation to drive homogeneous nucleation leading directly to graphite. The degree of supersaturation would be expected to be very large. Before this limit is reached, there are competing kinetic mechanisms tending to form soot through either dehydrogenation of acetylenic species or through polyaromatic hydrocarbons (PAHs) [27]. It is likely that these mechanisms would proceed rapidly enough to remove carbon from the gas phase before the critical level of supersaturation for homogeneous nucleation of graphite is reached. Thus the carbon should appear as soot rather than graphite as predicted in the present equilibrium calculations.

This example illustrates the importance of including all possible species, and their respective thermodynamic data, in the species set, and further illustrates the limitations of equilibrium modelling of systems in which deviations from equilibrium due to kinetic processes are expected to be significant.

An indication of which of the two mechanisms mentioned above is responsible for soot formation in a plasma reactor can be deduced from the discussion of Yoshihara and Ikegami [29] and knowledge of where the CH_4 is introduced into the plasma reactor. If the CH_4 is introduced upstream of the plasma, then the temperature at the reactor exit will be too high for PAH species to form and the high concentrations of C_2H_2 and C_2H predicted to occur in the gas phase suggest that dehydrogenation reactions of acetylenic species is the likely formation mechanism for soot. Conversely, if the CH_4 is introduced downstream of the plasma, then the cold CH_4 flow will be subject to rapid heating through the temperature window where PAH species are thought to form. Yoshihara and Ikegami argue that, while the temperature where sooting is occurring may be higher than the temperatures where PAH species are stable, the PAH species may form seed nuclei for subsequent soot particle growth as the temperature of the carbon containing species passes through the PAH species stability temperature window.

Despite the limitations of the data for solid carbon species, useful predictions can still be made using the data for graphite by assuming that if graphite is stable, then soot should also be a stable phase. This assumption should give an indication of the likely composition of the system at equilibrium with respect to condensed phase carbon species. Data presented by Yoshihara and Ikegami [27] for the Gibbs free energies of formation of various larger carbon clusters indicate that the deviation in free energies for these clusters from that of graphite, especially at higher temperatures, is minimal. Thus data for graphite can justifiably be used. A qualitatively more correct method to predict the behaviour of the system would be to kinetically model the system and work is currently underway within this group on this topic.

From the figures presented, it can be concluded that a ratio of 100 moles of argon per mole of TiO_2 is roughly equivalent to synthesis at a reduced pressure of 0.1 atmosphere with no argon in the system, while a ratio of 1000 moles of argon to 1 mole of TiO_2 is roughly equivalent to operating in an argon-free environment at a reduced system pressure of 0.01 atmospheres. However, the influence of the inert gas does not correspond to simply lowering the partial pressures of the reactants.

The results clearly demonstrate that any inert gases introduced into the system must be included in the system specified for thermochemistry equilibrium studies. The importance of including inert gases on the system specified for equilibrium studies was also briefly raised by Gans and Gauvin [28] in their Table 2, where they stated that the inert gas must be included in the system specification. It is significant that a large number of free energy minimisation analyses that have appeared in the literature appear to have neglected the effects of inert gas dilution [7-10, 26, 29-33].

Summary

The analyses presented in this manuscript have shown that TiC synthesis from TiO_2 and CH_4 in a thermal plasma torch is thermodynamically feasible within process condition windows identified by the analyses. The effects of the various process conditions on the expected condensed phases do not vary widely over the inert gas to TiO_2 molar ratios or the system pressure. For CH_4 to TiO_2 molar ratios from just above 1.0 to approximately 1.7, the predicted products are TiO with trace amounts of TiC, provided the cooling rates are high enough to prevent condensed-phase reactions. From a molar ratio of 1.7 to 3.0, the predicted products are TiC with trace amounts of TiO while above molar ratios of 3, the predicted products are TiC and soot.

To maximise the yield of TiC, non-equilibrium of the condensed-phase species compositions must be induced by rapidly quenching the system from the vicinity of 2000 K as this prevents gas-solid and gas-liquid reactions occurring. It has been argued that for systems containing excess carbon, soot formation will occur rather than graphite formation. To achieve better predictions of kinetically-controlled processes such as soot formation than is possible with thermochemistry equilibrium modelling, kinetic modelling should be used, or at least some estimates of soot formation kinetics must be employed.

The effects of inert gas dilution has been incorporated in the analyses presented and it was clearly demonstrated that any inert gases introduced into the system must be included in the system specified for thermochemistry equilibrium studies. The presence of an inert gas has an influence on the system similar to lowering the overall system pressure, but the effect is not the same as a simple lowering of the system partial pressures of the reactants so it must be explicitly included in the calculations.

Acknowledgements

This work has been supported by DSIR Chemistry under contract No. UV/CD/3/1.

References

[1] W.E. Kuhn, "The Formation of Silicon Carbide in the Electric Arc", J. Electrochem. Soc., **110**, 4, 298-306 (1963)

[2] J.D. Holmgren, J.O. Gibson and C. Sheer, "Some Characteristics of Arc Vaporized Submicron Particles", J. Electrochem. Soc., **111**, 3, 362-369 (1964)

[3] I.H. Warren and H. Shimzu, "Applications Of Plasma Technology In Extractive Metallurgy", Can. Min. Metall. Bull., **68**, 169-178 (1965)

[4] S.R. Blackburn, T.A. Egerton and A.G. Jones, "Vapour Phase Synthesis of Nitride Ceramic Powders Using a D.C. Plasma", Paper 47, Fine Ceramic Powder Conference, Warick, April (1990)

[5] J.S. McFeaters and J.J. Moore, "Application of Nonequilibrium Gas-Dynamic Techniques to the Plasma Synthesis of Ceramic Powders", in Z.A. Munir and J.B. Holt (eds.), "Application of Nonequilibrium Gas Dynamic Techniques to the Plasma Synthesis of Ceramic Powders", VCH Publ., New York, 431-446 (1990)

[6] S.V. Joshi, Q. Liang, J.Y. Park and J.A. Batdorf, "Effect of Quenching Conditions on Particle Formation and Growth in Thermal Plasma Synthesis of Fine Powders", Plasma Chem. Plasma Process., **10**, 2, 339-358 (1990)

[7] D.M. Coldwell, "A Thermodynamic Analysis of the Reduction of Silicon Oxides Using a Plasma", High Temp. Sci., **8**, 309-316 (1976)

[8] C.M. Wai and S.G Hutchison, "A Thermodynamic Study of the Carbothermic Reduction of Alumina in Plasma", Met. Trans. B, **21B**, 406-408 (1990)

[9] Y. Chang, R.M. Young and E. Pfender, "Thermochemistry Of Thermal Plasma Chemical Reactions. Part II. A Survey Of Synthesis Routes For Silicon Nitride Production", Plasma Chem. Plasma Process., **7**, 3, 299-316 (1987)

[10] P. Kong, T.T. Huang and E. Pfender, "Synthesis of Ultrafine Silicon Carbide Powders in Thermal Arc Plasmas", IEEE Trans. Plasma Sci., **PS-14**, 4, 357-369 (1986)

[11] A.G. Turnbull and M.W. Wadsley, "The CSIRO-SGTE THERMODATA System (Version V)", CSIRO, Inst. of Energy and Earth Resources, Division of Mineral Chemistry, Port Melbourne, Australia, (1987)

[12] C.W. Bale and G. Eriksson, "Metallurgical Thermochemical Databases - A Review", Can. Metallurg. Quart., **29**, 2, 105-132 (1990)

[13] E.K. Storms, "The Refractory Carbides", Vol. 2 in "Refractory Materials", J.L. Margraves (ed.), Academic Press, New York, (1967)

[14] K. Ishizaki, T. Egashira, K. Tanaka and P.B. Celis, "Direct Production of Ultra-fine Nitrides (Si_3N_4 and AlN) and Carbides (SiC, WC and TiC) Powders By the Arc Plasma Method", J. Mater. Sci., **24**, 3553-3559 (1989)

[15] E. Neuenschwander, "Herstellung und Charakterisierung von Ultrafeinen Karbiden, Nitriden und Metallen", J. Less-Common Met., **11**, 365-375 (1966)

[16] B. Mitrofanov, A. Mazza, E. Pfender, P. Ronsheim and L.E. Toth, "D.C. Arc Plasma Titanium and Vanadium Compound Synthesis from Metal Powders and Gas Phase Non-Metals", Mat. Sci. Eng., **48**, 21-26 (1981)

[17] L.E. Toth, "Transition Metal Carbides and Nitrides", Vol. 7 in "Refractory Materials", J.L. Margraves (ed.), Academic Press, New York, 1971

[18] G. Neumann, R. Keiffer and P. Ettmayer, "The System TiC-TiN-TiO", Monatshefte Für Chemie, **103**, 1130-1137 (1972)

[19] M.W. Chase, C.A. Davies, J.R. Downey, D.J. Frurip, R.A. McDonald and A.N. Syverud, "JANAF Thermochemical Tables", 3rd Ed., J. Phys. Chem. Ref. Data, Supplement No. 1, **14** (1985)

[20] J.S. McFeaters, "The Non-Equilibrium Gas Dynamic Synthesis Of Transition Metal Carbide Powders", PhD Thesis, Carnegie-Mellon University (1986)

[21] G. Herzberg, "Molecular Spectra and Molecular Structure: IV. Constants of Diatomic Molecules", 2nd ed., Van Nostrad, New York (1979).

[22] C.W. Bauschlicher Jr. and E.M. Siegbahn, "On the Low-Lying States of TiC", Chem. Phys. Lett., **104**, 4, 331-335 (1984)

[23] A.I. Kingon, L.J. Lutz and R.F. Davis, "Thermodynamic Calculations for the Chemical Vapour Deposition of Silicon Nitride", J. Amer. Ceram. Soc., **66**, 8, 551-558 (1983)

[24] A.I. Kingon, L.J. Lutz, P. Liaw and R.F. Davis, "Thermodynamic Calculations for the Chemical Vapour Deposition of Silicon Carbide", J. Amer. Ceram. Soc., **66**, 8, 558-566 (1983)

[25] S.L. Girshick and C-P Chiu, "Homogeneous Nucleation and Particle Growth in Thermal Plasma Synthesis", in Z.A. Munir and J.B. Holt (eds.), "Combustion and Plasma Synthesis of High-Temperature Materials", VCH Publ., New York, 349-356 (1990)

[26] Y. Chang and E. Pfender, "Thermochemistry of Thermal Plasma Chemical Reactions. Part I. General Rules for the Prediction of Products', Plasma Chem. Plasma Proc., **7**, 3, 275-297 (1987)

[27] Y. Yoshihara and M. Ikegami, "Homogeneous Nucleation Theory for Soot Formation", JSME Int. J., Series II, **32**, 2, 273-280 (1989)

[28] I. Gans and W.H. Gauvin, "The Plasma Production Of Ultrafine Silica Particles", Can. J. Chem. Eng., **66**, 438-444 (1988)

[29] Z.P. Lu, T.W. Or, L. Stachowicz, P. Kong and E. Pfender, "Synthesis of Zirconium Carbide in a Triple Torch Plasma Reactor Using Liquid Organometallic Zirconium Precursors", Mat. Res. Soc. Symp. Proc., **190**, 77-82 (1990)

[30] D. Degout, F. Kassabji and P. Fauchais, "Titanium Dioxide Plasma Treatment", Plasma Chem. Plasma Process., **4**, 3, 179-198 (1984)

[31] A. Huczko and P. Meubus, "RF Plasma Processing of Silica", Plasma Proc. Plasma Chem., **9**, 3, 371-386 (1989)

[32] T. Kameyama, K. Sakanaka, A. Motoe, T. Tsunoda, T. Nakanaga, N.I. Wakayama, H. Takeo and K. Fukuda, "Highly Efficient and Stable Radio-Frequency Thermal

Plasma System for the Production of Ultrafine and Ultrapure β-SiC Powder", J. Mat. Sci., **25**, 1058-1065 (1990)

[33] P.R. Taylor and S.A. Pirzada, "Synthesis of Ceramic Carbide Powders in a Non-Transferred Arc Thermal Plasma Reactor", in N. El-Kaddah (ed.), "Thermal Plasma Applications in Materials and Metallurgical Processing", TMS, 193-207 (1992)

SYNTHESIS OF ULTRA-FINE TITANIUM CARBIDE IN A NON-TRANSFERRED ARC THERMAL PLASMA REACTOR

Patrick R. Taylor, Shahid A. Pirzada and Thomas D. McColm

Department of Metallurgical Engineering, College of Mines
University of Idaho, Moscow, Idaho 83843

ABSTRACT

Titanium carbide is a hard, wear resistant, low density, high temperature material which exhibits excellent resistance to mechanical degradation. As a brittle material, its strength is dependant on microflaw size and hence particle size during sintering. A non-transferred arc, thermal plasma method for the generation of very fine (0.2 to 0.4 micron) sized titanium carbide powders directly from titanium dioxide and methane has been developed. The experimental system is described and the necessary operating conditions outlined and explained. The thermodynamic behavior of the system is also presented.

Plasma Synthesis and Processing of Materials
Edited by Kamleshwar Upadhya
The Minerals, Metals & Materials Society ©1993

Introduction

The principal properties of titanium carbide are high hardness, relatively low density and, above all, an exceptionally high melting point. This melting point, coupled with the reasonably high thermal conductivity, enables titanium carbide to perform in high temperature environments with minimal physical or mechanical degradation. Two uses that take advantage of this fact are protective coatings for rocket components and gas turbine blades [1]. The high melting point of titanium carbide is beneficial when considering potential applications, however, when considering synthesis and subsequent component formation the situation is reversed.

In addition to superior fabrication attributes, metallic components possess sufficient slip systems for dislocation motion to allow the absorption of energy at a micro-crack tip by plastic deformation and inhibit crack propagation. For titanium carbide, as is the case for the majority of ceramic materials, full density is unobtainable and there is no potential for dislocation motion to inhibit crack growth. i.e. Ceramics are brittle and therefore vulnerable to brittle fracture.

The crucial factor in determining the fracture strength of a ceramic component is maximum flaw size. Strength and modulus data pertaining to titanium carbide is obtained from resonance and ultrasonic tests on very fine, flaw free, single crystal whiskers [2]. In truth, no component actually exhibits these values.

It has been established, that control of maximum flaw size is the key to the control of ceramic mechanical performance because of their tendency to undergo brittle fracture. The most important factor in determining microstructural flaw size within a ceramic component is the size of the powder used at the sintering stage. The finer the powder, the smaller the maximum flaw size, the higher the mechanical performance of the component.

It is the above dependance of ceramic component performance on powder size that make the proposed plasma synthesis technique potentially useful. In one step the carbide will be both synthesized from a cheap and easily handled mineral (oxide such as rutile), and produced as a powder far finer than any existing production techniques can obtain. From this powder, ceramic components can be produced with a higher mechanical performance potential than has been achieved to date.

Synthesis of Titanium Carbide

The established method to produce carbide powders is the Acheson process [3]. Due to the necessity of milling the product in the Acheson process, ultra-fine powders are not possible. The levels of impurity are also relatively high.

Titanium carbide powder has been produced using plasma synthesis by L. R. Swaney [4], F.E.Groenig [5] and several others [6-8]. Throughout all the literature covering plasma synthesis of titanium carbide there exists a common theme. In most of the cases, the titanium source in the reactants is in the form of a gaseous titanium halide, predominantly one of the tetrahalides. This is reacted with a gaseous hydrocarbon, usually methane, in the presence of a reducing agent, usually hydrogen. The plasma reactors employed were generally RF or DC operated in non-transferred mode. All are laboratory curiosities and have very small production scales.

In this system titanium dioxide (TiO_2) and methane (CH_4) are fed into a plasma reactor. The theory behind the process is that the first section of the reactor is a vaporization zone, where all input species are vaporized or thermally dissociated into the various gases. As the flow of the plasma gases carry the vaporized oxide and hydrocarbons down the reactor, and hence down the temperature gradient, a region is reached where the solid carbide becomes thermodynamically favorable. This region can be referred to as the reaction zone. In the reaction zone it is desired that by one, or a series of chemical reactions, solid carbide seeds are precipitated from the gaseous reactants. The resultant seeds are carried down the remainder of the reaction chamber where they undergo a small amount of growth before entering a quench chamber which inhibits reoxidation and any further growth.

System Thermodynamics

The following thermodynamic plots were all created using data acquired from the CSIRO thermodynamic software package [9]. The principal on which the plots are based is system free energy minimization.

Figure 1 shows the thermodynamic equilibrium diagram for an input of 1 mole of solid titanium dioxide. At lower temperatures, it exhibits the liberation of O_2 gas and the simultaneous formation of solid Ti_4O_7. This configuration remains stable until 3850 K. At this temperature the oxide rapidly dissociates into three gaseous species; TiO, atomic oxygen and TiO_2. The gaseous titanium dioxide further dissociates into the first two species. It is these two species, TiO(g) and O(g), that are the principal stable species in the plasma flame zone temperature range of interest. At extremely high temperatures, above 5000 K, all stable species are in the atomic or ionic form, this is the case for any thermodynamic system.

Figure 2 exhibits the free energy minimization for an input of one mole of methane over the temperature range shown. Up to 2500K the stable species are solid carbon and H_2(g). At this temperature the diatomic hydrogen begins to atomize and the hydrocarbons C_2H(g) and then C_2H_2(g) start to form. It is these species that are stable in the plasma flame zone temperature range of interest. Above 4500 K the gaseous species atomic carbon and hydrogen become the stable phases. Figure 3 was obtained from input species of titanium dioxide and methane in the ratio 1:3, this is the stoichiometric ratio for the overall reaction given by:

$$TiO_2(s) + 3CH_4(g) = TiC(s) + 2CO(g) + 6H_2(g) \qquad (1)$$

At 5000K, which is of the order of the temperature inside a

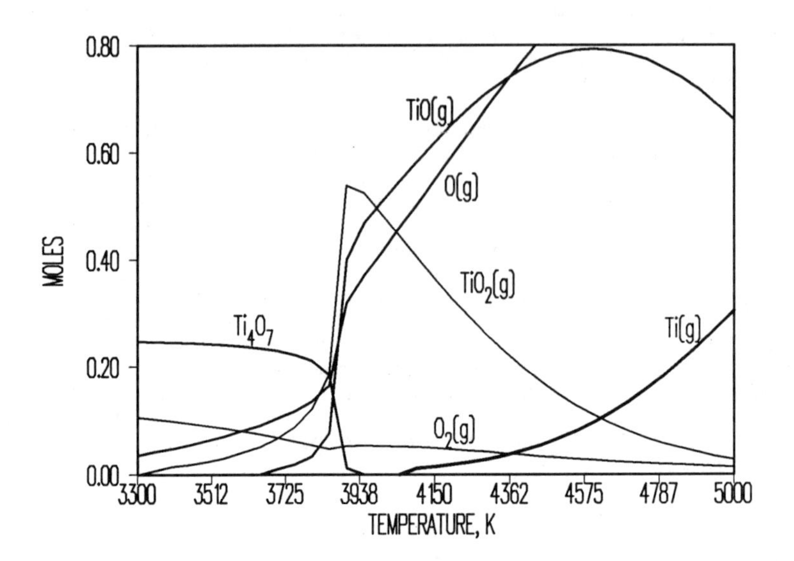

Figure 1 - Free Energy Minimization Plot for Titanium Dioxide.

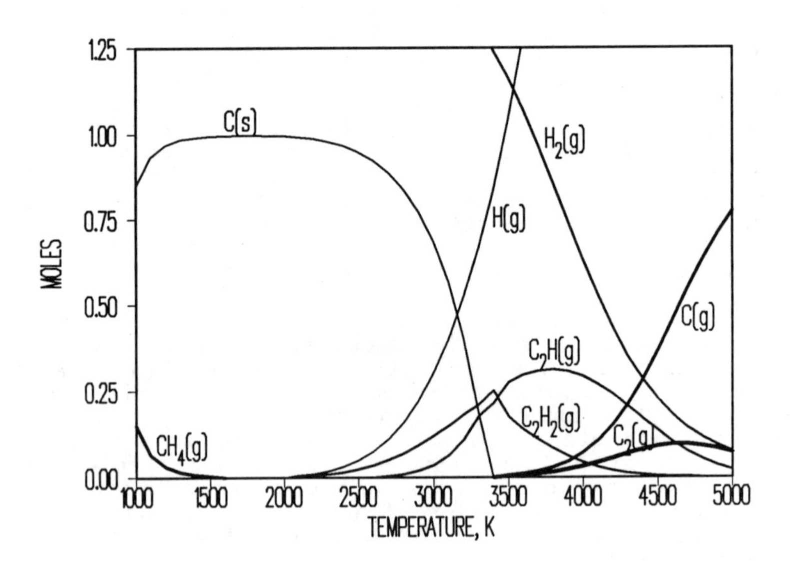

Figure 2 - Free Energy Minimization Plot for Methane.

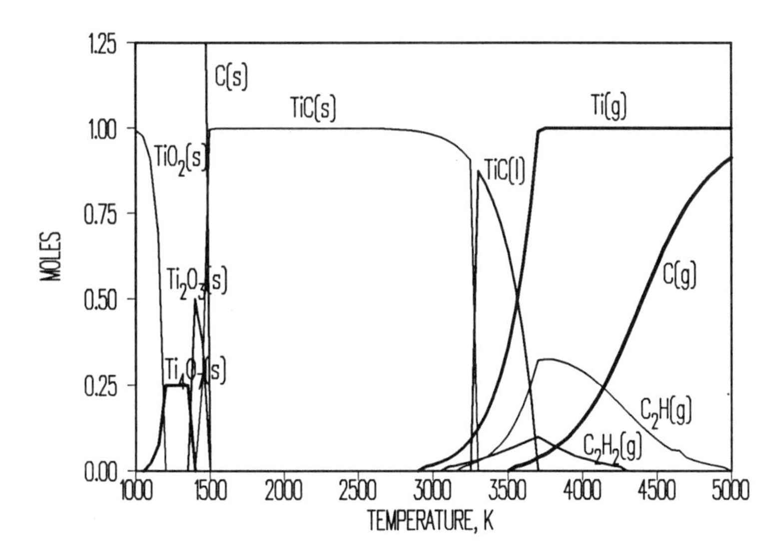

Figure 3 - Free Energy Minimization Plot for Methane:Titanium Dioxide (Molar Ratio 3:1).

plasma jet, only the monatomic gases Ti and C are stable. Mono atomic oxygen and hydrogen are also stable but are in too high a concentration to feature on the graph. At these temperatures, any chemical bond is broken by such intense external energy input. As the temperature gradient is descended, and hence the reactor traversed, the species begin to recombine. In the temperature range between 4500K - 3600K, titanium gas remains stable, however, carbon combines with hydrogen to form C_2H and C_2H_2 and with oxygen to form carbon monoxide. At this juncture it is necessary to point out that for all the above three plots carbon monoxide and diatomic hydrogen are consistently present across the majority of the temperature range but are in concentrations too high to be accommodated on a plot scaled to examine carbide formation. Below 3600K, the region encountered thermodynamically verifies the feasibility of the proposed process, titanium carbide becomes the predominant stable phase. There is a small region of liquid stability in the range 3600K - 3200K, below this, in concurrence with the stoichiometry of equation 1, all titanium present in the system is stable as 1 mole of solid titanium carbide until 1400K. At this temperature the oxides of titanium become stable.

Proposed Synthesis Route

Figure 4 exhibits the free energy change of reaction for the four principal oxide: hydrocarbon possibilities. All are negative across the entire temperature range, and hence all are possible. Due to stoichiometric considerations, and the information obtained from figures 2, 3 and 4 the reaction between TiO(g) and C_2H(g) is selected as the principle carbide forming

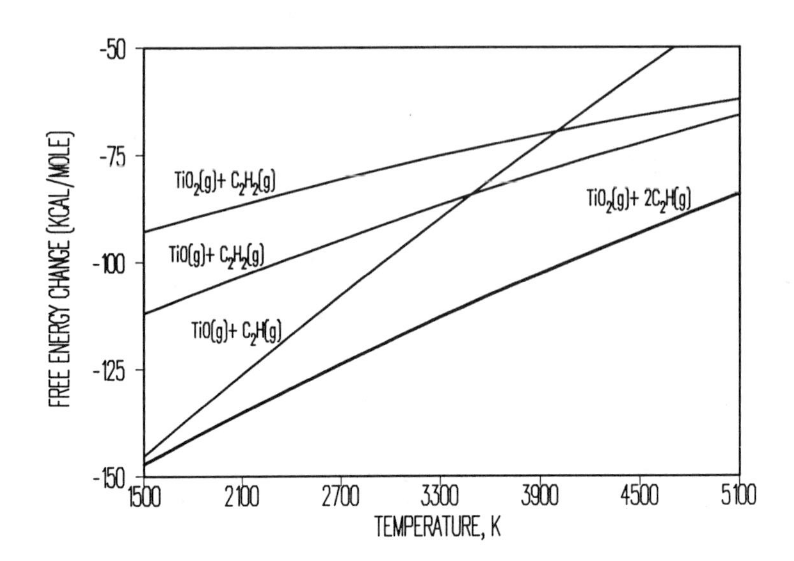

Figure 4 - Standard Free Energy for Various Reactions versus Temperature.

reaction. The proposed full synthesis sequence is detailed hereafter.

On injection into the vaporization zone titanium dioxide thermally decomposes and liberates diatomic oxygen according to equation 2:

$$4TiO_2(s) = Ti_4O_7(s) + 0.5O_2(g) \quad (2)$$

At the start of the reaction zone the suboxide has been resident sufficient time to acquire the necessary energy to thermally dissociate resulting in the formation of gaseous titanium monoxide and elemental oxygen. This process is described by equations 3 and 4:

$$Ti_4O_7(s) = 3TiO_2(g) + TiO(g) \quad (3)$$

$$3TiO_2(g) = 3TiO(g) + 3O(g) \quad (4)$$

Upon injection, methane is cracked instantly and the predominant stable hydrocarbon present is $C_2H(g)$, this interacts with the gaseous titanium monoxide in the reaction zone to produce solid titanium carbide according to equation 5:

$$TiO(g) + C_2H(g) = TiC(s) + CO(g) + 0.5H_2(g) \quad (5)$$

Experimental System

Figure 5 is a full schematic of the system. The central component of the experimental set up was the plasma reactor, which is a torch and chamber arrangement. The reaction chamber was a hollow graphite tube two or three feet in length. The internal diameter of the graphite tube was 3.5". The outer casing of the chamber was a double lined 316 stainless steel shell. The gap between the two walls of the steel shell was 0.5" and the necessary cooling water to prevent the reactor from over heating was pumped through the resultant hollow, cylindrical chamber. The gap in between the inner graphite tube and the outer steel shell was packed with zirconia felt. The zirconia felt was employed to insulate the graphite tube by reducing energy loss through the wall.

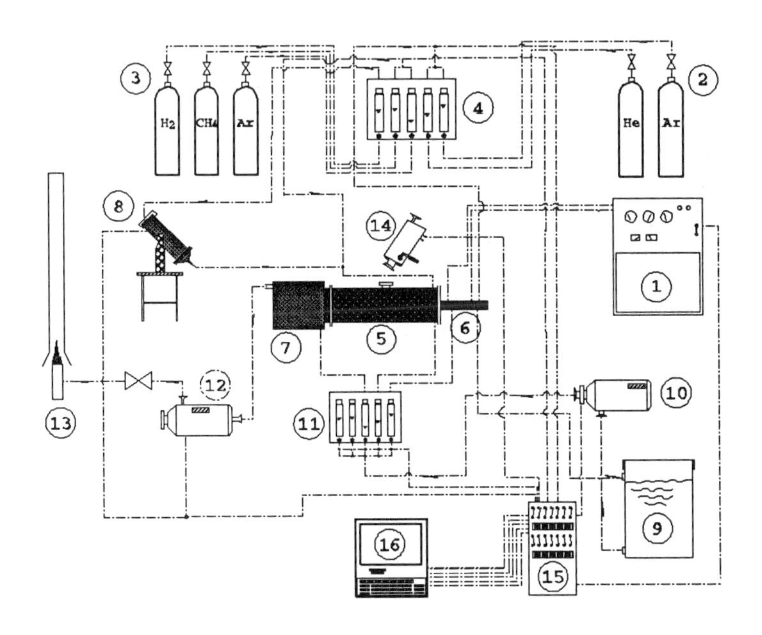

Figure 5 - Schematic of the Experimental System. (1-Power supply, 2-Plasma gases, 3-Carrier gases, 4-Gas flow meters, 5-Reactor, 6-Plasma torch, 7-Quench chamber, 8-Powder feeder, 9-Water tank, 10-Pump, 11-Water flow meters, 12-Filter, 13-Burner, 14-Pyrometer, 15-Data acquisition, 16-Computer)

The exit end of the tubular section was bolted to a one cubic foot quench chamber. It was within this chamber that the synthesized powder was quenched and collected. The rapid cooling within the quench chamber was due to both expansion and intensive water circulation. The water was pumped through a vertical baffle plate and a copper coil positioned to the rear of the plate. The two cooling elements were attached to the removable upper plate of the chamber. During operation the upper plate was bolted and clamped in position, teflon seals were employed to ensure that there was no leakage around the plate. The plasma torch was inserted into the reacting chamber at the open end through the center of a stainless steel plate and a graphite ring, it was stabilized by means of a stainless steel joint union which also prevented backward gas leakage at the point of insertion.

Experimental Procedure

The plasma gas was always a combination of two gases, the various gases experimented with were argon, helium, methane and hydrogen. A propane burner was used to burn off potentially flammable gas after it exited the reactor through the filter. Initial current and voltage levels were set at 120 V and 150 A and the arc was struck. Within the first few minutes of running, the gas flow rate and current were increased until the desired operating power was attained. During the initial heat up period the various gas and water temperatures were continually monitored by means of a series of thermocouples positioned at key points along the reactor. The read out from these thermocouples was displayed both on the system control panel and the screen of a computer data acquisition system. The system was heated up until a steady state was attained.

At various points in the first one foot section of the reactor, small powder injection ports were drilled and lined with .25" hollow graphite tubes. Carrier gas was channelled from the gas cylinders into the powder feeder. Previous to running the experiment the powder feeder was calibrated so that the mass feed rate was known, and hence, the correct volumetric flow rate of methane could be selected to give the desired stoichiometric reaction ratio. A mass feed rate of three grams per minute was found to be most suitable.

Argon, helium and methane were employed as carrier gases, the carrier gas for powder feeder was usually a combination of two of the three. Titanium dioxide (rutile) powder was picked up in the feeder by the carrier gas and carried through tubing into the graphite injection tube, and hence into the reactor. The powder giving the most satisfactory feed characteristics was <325 mesh, commercial grade, titanium dioxide. The feeder was run for a period of 15 to 20 minutes during which time the temperatures were continually monitored. After the system was totally shut down and cooled, the collection chamber opened and experimental product removed. Table 1 summarizes the sequence of experimental variables that were evaluated and that led to the establishment of a successful synthesis procedure.

Results

An important aspect of the research was to establish the precise chemical and physical nature of the material produced and verify the authenticity of the process. The analysis techniques used were predominantly standard. The fineness of powders produced made X-ray powder diffractometry an excellent method of crystal structure determination. Scanning electron microscopy was employed on the feed and product powders. This was purely to obtain information on particle size and morphology, and in the case of the product powder, an independent estimate of the size distribution. A simple roast and leach procedure was devised and performed to deduce the percentage of free carbon and un/re-oxidized powder within the final product.

After the product powders were collected from the plasma reactor they were subjected to a low temperature, controlled oxygen roast to remove the excess carbon as carbon dioxide. Only

Table 1 - Experimental Variables Studied

RUN #	PLASMA GAS	CARRIER GAS l/min	FEED PORT *	FEED RATE g/min
TIT 1	Ar:He	Ar:CH4	A	11
TIT 2	Ar:He	Ar:CH4	A	5
TIT 3	Ar:He	Ar:CH4	A	5
TIT 4	Ar:He	Ar:CH4	B	5
TIT 5	Ar:He	Ar:CH4	B	5
TIT 6	Ar:He	Ar:CH4	B	5
TIT 7	Ar:He	Ar:CH4	B	5
TIT 8	Ar:CH4	Ar	A	6.5
TIT 9	Ar:H2	Ar:CH4	C	5
TIT 10	Ar:H2	Ar:CH4	C	5
TIT 11	Ar:H2	He:CH4	C	3
TIT 12	Ar:H2	He:CH4	C	3
TIT 13	Ar:H2	He:CH4	C	3

*
Feed port locations:
Feed port A: Three inches down first section, perpendicular to plasma gas flow through the side of the cylindrical lining.
Feed port B: Zero inches down first section, parallel with plasma gas flow through steel and graphite end plates.
Feed port C: Nine inches down first section, perpendicular to plasma gas flow through the top of the cylindrical lining.

a small percentage of the titanium carbide was oxidized. The roasted powder was then subjected to a heated sulfuric acid leach to remove the residual titanium oxides. This procedure resulted in a nearly pure and fine sized titanium carbide powder.

Figure 6 and 7 are SEM micrographs of the feed oxide and final product powder respectively. The success of producing fine powders from course feed oxides is excellently displayed. Figures 8 and 9 exhibit the size distribution for the same powders, these were determined by a centrifugal sedigraph method for the produced powder and Coulter Counter for the feed powder.

Conclusions

Submicron crystalline titanium carbide powder was directly synthesized in a non-transferred arc thermal plasma reactor from mineral titanium dioxide and methane. The conversion of oxide fed to carbide collected was approximately 80% for the best experiment to date. The purity of the powder was greater than 99% after treatment. The size of the product powder was between 0.2and 0.4 micron. The system and procedure to perform the synthesis were standardized and documented. The thermodynamics of the system were thoroughly analyzed and a synthesis route was proposed.

Acknowledgement

This research was supported by the National Science Foundation under grant RII-8902065.

Figure 6 - SEM Micrograph of the Feed Titanium Dioxide Powder

Figure 7 - SEM Micrograph of the Product Titanium Carbide Powder.

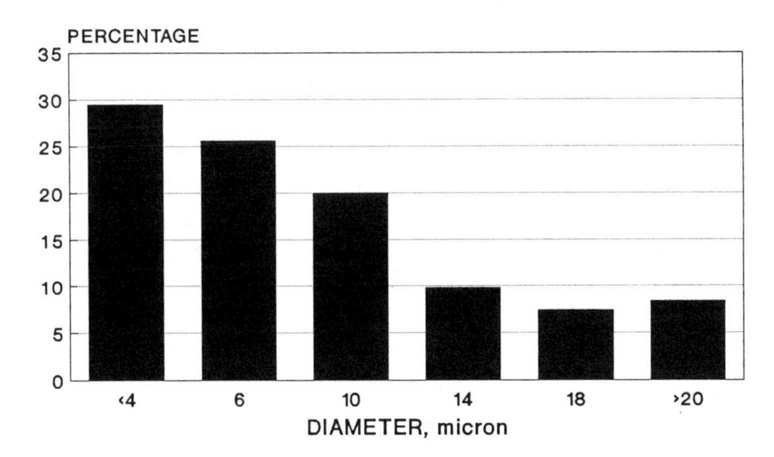

Figure 8 - Particle Size Distribution of the Feed Titanium Dioxide Powder.

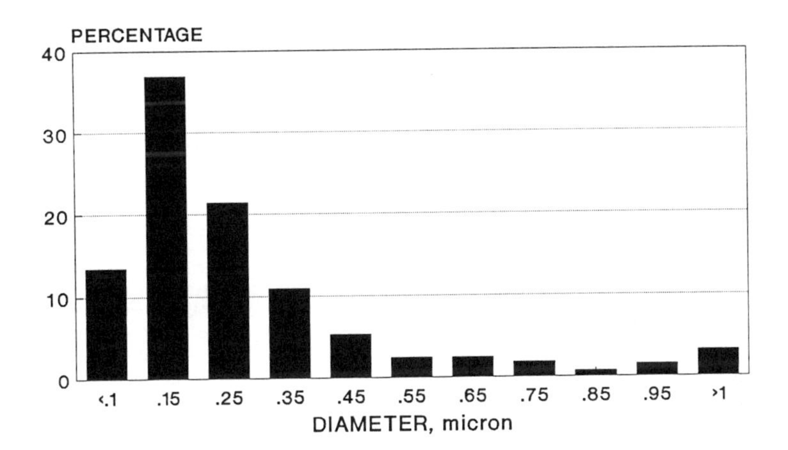

Figure 9 - Particle Size Distribution for the Product Titanium Carbide Powder.

References

1. I.J. McColm, "Ceramic Science", Blackie, Glasgow, U.K., (1983), 320.

2. J.B. Wachtman Jr., "Determination of Elastic Constants Required for Application of Fracture Mechanics to Ceramics, "Fracture Mechanics of Ceramics", Plenum Press, NY, vol 1, (1974), 49-68.

3. I.J. McColm and N.J. Clark, "High Performance Ceramics", Blackie, Glasgow, U.K. (1988), 99-100.

4. L.R. Swaney, "Preparation of Submicron Titanium Carbide", U.S. Patent 3485586, (1969).

5. F.E. Groenig, "Preparation of Titanium Carbide", U.S. Patent 3761576, (1973).

6. G. Perugini, "Arc-Plasma Reactions for Ceramics", Comm. 3rd Symp. Int. Chem. Plasma, Papers S.3.7, S.4.4 & S.4.5, (1977).

7. A.J. Becker et al., "Plasma Process Routes to Synthesis of Carbide, Boride and Nitride Ceramic Powders", Materials Research Symposium Proceedings, 98, (1987), 335-346.

8. P.C. Kong and Y.C. Lau, "Plasma Synthesis of Ceramic Powders", Pure and Applied Chemistry, 62(9), (1990), 1809-1816.

9. A.G. Turnbull and M.W. Wadsley," The CSIRO Thermochemistry System, version 5, IMEC, Australia, (1988).

KINETIC MODELING OF TITANIUM CARBIDE SYNTHESIS IN

THERMAL PLASMA REACTORS

John S. McFeaters, Robert L. Stephens and Peter Schwerdtfeger

Computational Materials Science and Engineering
Research Unit
School of Engineering
Auckland University
Private Bag 92019
Auckland, New Zealand

Abstract

Experimental work using thermal plasma reactors to produce titanium carbide ceramic powders has shown that there is competition between product formation and soot formation mechanisms. In order to better understand this process and improve product quality, a numerical model has been developed to study the kinetic behavior of this process. The model incorporates chemical kinetics, nucleation and growth, and soot formation mechanisms.

The chemical kinetic model was based on acetylene and ethene pyrolysis, and methane combustion, with the addition of a number of reactions to account for the free carbon species found at the high temperatures in thermal plasma reactors. A system of reactions describing the chemistry of the titanium based molecules (Ti, TiC, TiO and TiO_2) with the other reactants was included and reaction rates were estimated using standard techniques. Nucleation was modeled using the capillarity approximation and soot formation was based on a simple model that assumes that once carbon begins to polymerize, it goes irreversibly to soot. The governing equations were integrated through time using typical time-temperature histories found by computational fluid dynamics (CFD) modeling of an RF plasma torch.

The results indicate that there is a delicate balance between system pressure, cooling rates, product formation and soot formation, and that this balance may be a limiting feature of ceramic carbide production in thermal plasma reactors.

Plasma Synthesis and Processing of Materials
Edited by Kamleshwar Upadhya
The Minerals, Metals & Materials Society ©1993

Introduction

Thermal plasma reactors are used commercially for the synthesis of a number of different ultra-fine ceramic powders [1,2]. Thermal plasma processing offers potential advantages over other methods with the ability to generate ultra-fine powders and metastable stoichiometries [1-4].

Our group is currently developing thermal plasma technology for the synthesis of high specification titanium based ceramic materials. Ideally in this process, a source of titanium and carbon are fed into the system and titanium carbide (TiC) is formed in the gas phase as the hot reactant stream leaves the plasma reactor and begins to cool. The titanium carbide then homogeneously nucleates and the nuclei grow by condensation to form product particles in the size range of 0.01 to 0.1 μm that are recovered downstream [5,6]. In practice however, experiments on the synthesis of titanium carbide powders have shown that the product quality is often compromised by the formation of soot during the process. This competition between the formation of titanium carbide and the formation of soot is observed during the direct reduction of titania (TiO_2) powders to TiC as well as with the use titanium (Ti) powder feedstock for TiC production. In both cases pure methane or natural gas has been used as the source of carbon. Experimental observations indicate that this behavior occurs frequently in the processing of other metal carbide powders as well [3,6-10]. In order to better understand this complex process, a numerical model was developed to take into account chemical kinetics, nucleation and condensation, and soot formation.

Thermal plasma reactors used for ceramic powder synthesis generally use either a radio frequency (RF) inductively coupled power supply or a direct current (DC) electric discharge to supply power to the plasma. Peak temperatures are in the order of 10,000 K for RF plasma torches and 15,000 K for DC plasma torches and pressures are in the order of 1 to 10 atm. Under these conditions, local thermodynamic equilibrium (LTE) may be assumed. Typical residence times in the plasma fireball are in the order of 1 to 10 ms. Reactants passing through this high temperature region are essentially broken down into their component atoms, ions and electrons. The high temperatures in the plasma are necessary in order to provide sufficient electrical conductivity to sustain the plasma and to vaporize the solid phase reactants.

As the reactant stream leaves the plasma fireball and begins to cool, the process is characterized by high temperatures and steep gradients in temperature and species concentration. Temperatures drop from the plasma temperature down to the order of 1000 K over a period of 10 to 100 ms. The rate of cooling is typically in the order of 10^4 to 10^6 K/s. However, gas dynamic techniques and mixing with cold gas streams can enhance this to the order of 10^8 K/s [11,12]. During this cooling period rapid gas phase chemical reactions occur forming the product precursors and finally the products. As the temperature drops and product molecules begin to form at a rapid rate, the reactant stream can become highly supersaturated. The degree of supersaturation can be high enough to drive a homogeneous nucleation. The subsequent growth of the clusters can lead to the formation of ultra fine particles. The extreme temperatures, rapid cooling and steep gradients create unusual process conditions which, in some circumstances, can give rise to significant deviations from kinetic equilibrium and chemical equilibrium [13-15].

Experiments have shown that carbon in the system can condense as soot which is undesirable since it is difficult to separate the TiC particles from the soot particles. Previous equilibrium modeling work [16] has shown that there is a balance between product formation, sooting and reformation of the oxide.

Often, as is the case here, the desired products are only metastable in the system. TiC is the thermodynamically favored state at higher temperatures. However, when the system temperature falls below ~ 1500 K, the metal oxide again becomes thermodynamically favored. In order to produce TiC in a continuous one step reduction-product formation process, the product must be chemically formed and condensed before the system temperature falls to low levels. Once the product is in the condensed phase, loss of the product is minimal because cooling rates are still high and condensed phase reactions are relatively much slow. However, the loss of some product due to surface oxidation may be unavoidable unless special precautions are taken during handling of the product.

The Model

Assumptions

The following assumptions were made in setting up the model;

- Constant pressure, based on the relatively low gas velocities.

- Homogeneous composition: it is assumed that all solid phase reactant material is completely vaporized and that the system is perfectly mixed.

- Heat transfer and fluid mechanics are uncoupled from the chemistry, nucleation, condensation and soot processes. The time-temperature profiles were taken from CFD modeling of our plasma torch [17].

- Ideal gas behavior was assumed based on low densities and high temperatures.

- No ionized species were considered. The reactions of interest and product formation do not occur until the system temperature has dropped below 5000 K at which point there is no significant ionization. However, it should be noted that ionization may play an important role in the formation of soot in some circumstances [18].

- Reverse reaction rates for all reactions were found using equilibrium calculations and detailed balance.

Species

The system was restricted to the investigation of the kinetics found in the operating range of the torch where product formation occurs. This essentially eliminates many of the higher hydrocarbon species such as polycyclic aromatic hydrocarbons (PAH) as these species are not found above 3000 K [19], and while polyacetylenic chain species may be present in the system, these species usually react irreversibly towards soot and are implicitly accounted for in the soot formation mechanism used in this model.

Table I. List of Reactive Species Included in Kinetic Mechanism

O	O_2	CO	CO_2	H	H_2	OH	H_2O	HO_2
C	CH	CH_2	CH_3	CH_4	HCO	CH_2O	C_2	C_2O
C_2H	C_2H_2	C_2H_3	C_2H_4	C_3	Ti	TiC	TiO	TiO_2

The chemical system considered consisted of 27 reacting species plus an inert gas which in this case was argon. These species included Ti, TiC, TiO and TiO_2, plus the carbon species C, C_2 and C_3. The addition of the carbon molecules is somewhat of a departure from typical combustion and pyrolysis modeling. However, the time-temperature history here is markedly different from that of most combustion and pyrolysis processes. Typically, combustion and pyrolysis occurs in a temperature range of between approximately 1500 and 3500 K. Plasma processing is in effect combustion in reverse. The reactants are heated to temperatures in the order of 10,000 K. At these peak temperatures, the mixture consists of ions, electrons and atoms with virtually no molecules present in the system. Equilibrium is reached in the order of several microseconds and no further reaction occurs until the reactant stream begins to cool after leaving the plasma fireball. During this cooling period, the bulk of the product formation occurs in the temperature range from 5000K down to around 2500 K. As the reactant stream is cooled through this range of temperatures, almost all of the carbon becomes bound in the form of acetylene, TiC or soot, and virtually all the oxygen is irreversibly bound in CO. It is therefore assumed that at the lower temperatures there is no carbon or oxygen available for the formation of more complex hydrocarbon species and we eliminated them from consideration. We have considered species containing up to 2 carbon atoms as well as the C_3 molecule. The reactive species considered are given in Table I.

Table II - Estimated Gas Phase Molecular Properties of Titanium Carbide, TiC(g).

mol. wt. = 59.91 gm/mol	$\Delta H_f^o(0\ K)$ = 648.5 kJ/mol	D_o = 531.4 kJ/mol
ω_e = 1367 K	B_e = 0.8046 K	σ = 1
g_o = 3	g_1 = 1	ε_1 = 618.7 K
$T_{boiling}$ = 4986 K (for congruent vaporization)		$\Delta H_{vap}(T_b)$ = 664.7 kJ/mol

Thermodynamics

Thermodynamic properties of the molecules were calculated using statistical mechanical techniques. Data for all the species were taken from the JANAF tables [20] with the exception of TiC(g). The properties of TiC(g), Table II, were estimated from spectroscopic data and data from *ab initio* molecular orbital configuration interaction calculations [21-24].

Table III - Elementary reactions included in the kinetic mechanism.

	Reactions									Forward rate coefficients[a] A (exp)	n	E_a(K)	$\Delta H_{f\text{-}298}$	ref
1	H	+ O_2	↔ OH	+ O						1.59 (+11)	-0.927	8500	70.51	25
2	O	+ H_2	↔ OH	+ H						3.87 (-02)	2.70	3150	8.17	25
3	OH	+ H_2	↔ H_2O	+ H						2.16 (+02)	1.51	1730	-63.16	25
4	OH	+ OH	↔ O	+ H_2O						2.10 (+02)	1.40	-190	-71.33	25
5	O	+ O	+ M	↔ O_2	+ M					1.00 (+11)	-1.00		-498.34	25[b]
6	H	+ H	+ M	↔ H_2	+ M					6.40 (+11)	-1.00		-436.00	25[b]
7	H	+ OH	+ M	↔ H_2O	+ M					8.40 (+15)	-2.00		-499.16	25[b]
8	H	+ O_2	+ M	↔ HO_2	+ M					7.00 (+11)	-0.80		-205.45	25[b]
9	HO_2	+ H	↔ OH	+ OH						1.50 (+08)		500	-151.87	25
10	HO_2	+ H	↔ H_2	+ O_2						2.50 (+07)		350	-230.55	25
11	HO_2	+ H	↔ H_2O	+ O						5.00 (+06)		710	-223.20	25
12	CO	+ OH	↔ CO_2	+ H						1.17 (+01)	1.35	-360	-104.31	25
13	CO	+ HO_2	↔ CO_2	+ OH						1.50 (+08)		11870	-256.18	25
14	CO	+ O	+ M	↔ CO_2	+ M					3.01 (+08)		1515	-532.14	25[b]
15	CO	+ O_2	↔ CO_2	+ O						2.50 (+06)		24050	-33.80	25
16	CH	+ OH	↔ HCO	+ H						3.00 (+07)			-376.69	25
17	CH	+ O_2	↔ CO	+ OH						4.47 (+05)	0.5		-668.54	25
18	HCO	+ M	↔ H	+ CO	+ M					1.86 (+11)	-1.00	8550	65.47	25[b]
19	CH_2	+ OH	↔ CH	+ H_2O						1.13 (+01)	2.00	1515	-74.22	25
20	CH_2	+ O_2	↔ CO_2	+ H	+ H					1.60 (+06)		500	-347.91	25
21	CH_2	+ O_2	↔ CH_2O	+ O						5.00 (+07)		4535	-249.82	25
22	CH_2	+ O_2	↔ CO_2	+ H_2						6.90 (+05)		2525	-783.91	25
23	CH_2	+ O_2	↔ CO	+ OH	+ H					8.60 (+04)		-2525	-243.60	25
24	CH_2	+ O_2	↔ HCO	+ OH						4.30 (+04)		-2525	-309.07	25

Kinetic Mechanism

There is currently very little known about gas phase kinetics involving titanium compounds. There is no data available for reactions involving Ti atoms in the gas phase and very little information available on high temperature reactions involving free carbon species that were thought to be important in this scheme.

Table III - Continued.

	Reactions	A (exp)	n	E_a(K)	$\Delta H_{f\text{-}298}$	ref
		Forward rate coefficients[a]				
25	$CH_2 + CH_2 \leftrightarrow C_2H_2 + H + H$	1.00 (+08)			-117.42	25
26	$CH_2O + H \leftrightarrow HCO + H_2$	1.26 (+02)	1.62	1100	-67.42	25
27	$CH_2O + O \leftrightarrow HCO + OH$	3.50 (+07)		1770	-59.25	25
28	$CH_3 + OH \leftrightarrow CH_2 + H_2O$	1.13 (+00)	2.13	1225	-37.61	25
29	$CH_3 + O_2 \leftrightarrow CH_2O + OH$	5.20 (+07)		17560	-216.10	25
30	$CH_3 + CH_3 \leftrightarrow CH_4 + CH_2$	1.70 (+03)	0.56	6325	21.81	25
31	$CH_4 + M \leftrightarrow CH_3 + H + M$	1.29 (+27)	-3.73	53594	439.74	25
32	$CH_4 + H \leftrightarrow CH_3 + H_2$	3.90 (+00)	2.11	3900	3.74	25
33	$CH_4 + O \leftrightarrow CH_3 + OH$	1.90 (+03)	1.44	4365	11.91	25
34	$CH_4 + O_2 \leftrightarrow CH_3 + HO_2$	8.00 (+07)		28180	234.29	25
35	$CH_4 + OH \leftrightarrow CH_3 + H_2O$	1.50 (+00)	2.13	1225	-59.42	25
36	$C_2H + H_2 \leftrightarrow C_2H_2 + H$	1.10 (+07)		1443	-89.11	25
37	$C_2H_2 + O \leftrightarrow CH_2 + CO$	7.81 (-03)	2.80	250	-196.69	25
38	$C_2H_2 + OH \leftrightarrow C_2H + H_2O$	2.71 (+07)		5280	25.95	25
39	$C_2H_2 + OH \leftrightarrow CH_3 + CO$	4.85 (-10)	4.00	-1010	-230.41	25
40	$C_2H_2 + M \leftrightarrow C_2H + H + M$	4.17 (+10)		53880	525.11	25[b]
41	$C_2H_3 + O_2 \leftrightarrow CH_2O + HCO$	4.00 (+06)		-120	-346.07	25
42	$C_2H_3 + M \leftrightarrow C_2H_2 + H + M$	2.00 (+32)	-7.17	25460	165.91	25[b]
43	$C_2H_4 + H \leftrightarrow C_2H_3 + H_2$	3.16 (+05)	0.70	4030	9.19	25
44	$C_2H_4 + O \leftrightarrow CH_3 + HCO$	1.60 (+02)	1.44	260	-112.61	25
45	$C_2H_4 + O \leftrightarrow CH_2O + CH_2$	7.11 (+02)	1.55	975	-19.64	25
46	$C_2H_4 + OH \leftrightarrow C_2H_3 + H_2O$	3.00 (+07)		1503	-53.97	25
47	$C_2H_4 + CH_3 \leftrightarrow C_2H_3 + CH_4$	3.92 (+06)		6555	5.45	25
48	$C_2H_4 + M \leftrightarrow C_2H_2 + H_2 + M$	2.60 (+11)		39890	175.10	25[b]
49	$C_2H_4 + M \leftrightarrow C_2H_3 + H + M$	3.80 (+11)		49400	445.19	25[b]
50	$HO_2 + O \leftrightarrow OH + O_2$	1.75 (+07)		-200	-222.38	41
51	$HO_2 + OH \leftrightarrow H_2O + O_2$	1.45 (+10)	-1.00		-293.71	41
52	$O + CH \leftrightarrow H + CO$	6.14 (+07)		914	-739.05	42
53	$HCO + O_2 \leftrightarrow CO + HO_2$	5.12 (+07)		850	-139.98	41
54	$CH_2O + OH \leftrightarrow HCO + H_2O$	3.00 (+07)		601	-130.58	43
55	$CH_2O + O_2 \leftrightarrow HCO + HO_2$	2.05 (+07)		19600	163.13	41
56	$CH_2O + M \leftrightarrow HCO + H + M$	1.20 (+35)	-6.90	48590	368.58	41[c]
57	$CH_2O + M \leftrightarrow CO + H_2 + M$	2.50 (+08)		14250	-1.95	44
58	$CH_3 + OH \leftrightarrow CH_2O + H_2$	3.19 (+06)	-0.53	5440	-294.78	45
59	$O + CH \leftrightarrow C + OH$	1.52 (+07)		2381	90.51	42
60	$CH_3 + M \leftrightarrow CH_2 + H + M$	1.00 (+10)		45584	461.55	43[c]
61	$CH_2 + CH_2 \leftrightarrow CH_3 + CH$	2.40 (+08)		5000	-36.61	46
62	$CH_3 + CH_2O \leftrightarrow CH_4 + HCO$	5.54 (-03)	2.81	2950	-71.16	41
63	$CH + CH_4 \leftrightarrow CH_2 + CH_3$	3.01 (+07)		-200	14.80	47
64	$CH + CO_2 \leftrightarrow CO + HCO$	3.44 (+06)		345	-272.38	48
65	$C + H_2 \leftrightarrow CH + H$	4.00 (+08)		11700	98.68	49
66	$C + O_2 \leftrightarrow CO + O$	1.20 (+08)		2010	-578.03	49
67	$C + C + M \leftrightarrow C_2 + M$	1.81 (+15)	-1.60		-595.64	50

The reactions included in the system were based primarily on acetylene and ethene pyrolysis, and methane combustion with the addition of a number of reactions to account for the gas phase molecules containing Ti. Initially, a computer program was written to generate all possible bimolecular reactions of the species. This list of 423 reactions was compared to the reaction

Table III - Continued.

	Reactions	Forward rate coefficients[a] A (exp)	n	E_a(K)	$\Delta H_{f\text{-}298}$	ref
68	$O + H + M \leftrightarrow OH + M$	4.71 (+06)	-1.00		-427.83	41[c]
69	$CH_2 + O \leftrightarrow CO + H_2$	5.00 (+06)		4125	-750.11	25[d]
70	$CH_3 + HCO \leftrightarrow CH_4 + CO$	3.20 (+05)	0.50		-374.27	25[e]
71	$HCO + H \leftrightarrow CO + H_2$	3.23 (+06)	0.5		-370.53	41[e]
72	$HCO + O \leftrightarrow CO + OH$	8.04 (+05)	0.5		-362.36	41[e]
73	$HCO + O \leftrightarrow H + CO_2$	8.04 (+05)	0.5		-466.67	41[e]
74	$HCO + OH \leftrightarrow CO + H_2O$	8.04 (+05)	0.5		-433.69	41[e]
75	$CH_2 + H \leftrightarrow CH + H_2$	4.36 (+06)	0.5		-11.06	41[e]
76	$CH_2 + O \leftrightarrow CO + H + H$	3.23 (+05)	0.5		-314.11	41[e]
77	$CH_2 + OH \leftrightarrow CH_2O + H$	4.83 (+05)	0.5		-320.33	41[e]
78	$CH_2 + H_2 \leftrightarrow CH_3 + H$	8.04 (+01)	0.5		-25.55	41[e]
79	$CH_3 + O \leftrightarrow CH_2O + H$	2.09 (+06)	0.5		-286.61	41[e]
80	$C_2H + O \leftrightarrow CH + CO$	4.84 (+05)	0.50		-296.86	41[e]
81	$C_2H_3 + H \leftrightarrow C_2H_2 + H_2$	2.57 (+06)	0.50		-270.09	41[e]
82	$C_2H_3 + OH \leftrightarrow C_2H_2 + H_2O$	8.04 (+05)	0.50		-333.25	41[e]
83	$CO_2 + CH_2 \leftrightarrow CH_2O + CO$	6.28 (+02)	0.50		-216.02	41[e]
84	$CH_2 + HCO \leftrightarrow CH_3 + CO$	4.83 (+05)	0.50		-396.08	41[e]
85	$CH_2 + CH_2O \leftrightarrow CH_3 + HCO$	1.61 (+02)	0.50		-92.97	41[e]
86	$CH_2 + CH_3 \leftrightarrow C_2H_4 + H$	1.13 (+06)	0.50		-266.97	41[e]
87	$CH_2 + CH_2 \leftrightarrow C_2H_2 + H_2$	2.25 (+05)	0.50		-533.42	51[e]
88	$CH_2 + CH_2 \leftrightarrow C_2H_3 + H$	4.53 (+05)	0.50		-283.33	46[e]
89	$C + CO + M \leftrightarrow C_2O + M$	1.32 (+09)	0.50		-319.54	52[c,e]
90	$HCO + HCO \leftrightarrow CH_2O + CO$	4.83 (+05)	0.5		-303.11	41[e]
91	$C_2O + O \leftrightarrow CO + CO$	6.85 (+06)	0.50	245	-756.83	53[e]
92	$CH_2 + M \leftrightarrow CH + H + M$	1.30 (+07)	0.50	50606	424.94	[f]
93	$O_2 + C_2O \leftrightarrow CO_2 + CO$	8.70 (+06)	0.50	1975	-790.63	54[e]
94	$C_2O + H \leftrightarrow CH + CO$	1.84 (+07)	0.50	950	-17.78	53[e]
95	$CH + OH \leftrightarrow C + H_2O$	5.00 (+03)	0.50		-161.84	55[e]
96	$C + CO_2 \leftrightarrow CO + CO$	7.50 (+04)	0.50	2335	-544.23	55[e]
97	$C_2 + H_2O \leftrightarrow C_2O + H_2$	7.00 (+06)	0.50	2645	-309.28	56[e]
98	$C_2 + CO_2 \leftrightarrow CO + C_2O$	1.00 (+07)	0.50	2750	-268.13	56[e]
99	$C + O + M \leftrightarrow CO + M$	2.40 (+04)		2184	-1076.37	57
100	$O + C_2 \leftrightarrow CO + C$	6.30 (+05)	0.50		-480.73	57
101	$O + C_3 \leftrightarrow CO + C_2$	6.30 (+05)	0.50	1030	-342.00	57[g]
102	$CH + M \leftrightarrow C + H + M$	9.73 (+06)	0.50	40560	337.32	[h]
103	$C + OH \leftrightarrow CO + H$	3.51 (+04)	0.50		-648.54	[h]
104	$OH + C_2 \leftrightarrow H + C_2O$	5.76 (+02)	0.50		-372.44	[h]
105	$OH + C_3 \leftrightarrow HCO + C_2$	8.93 (+03)	0.50		20.36	[h]
106	$C + CH \leftrightarrow C_2 + H$	4.06 (+04)	0.50		-258.32	[h]
107	$CH + CH \leftrightarrow C + CH_2$	5.71 (+03)	0.50		-87.62	[h]
108	$CH_2 + CH \leftrightarrow C + CH_3$	7.50 (+04)	0.50		-124.23	[h]
109	$C_3 + M \leftrightarrow C_2 + C + M$	1.39 (+07)	0.50	87670	734.37	[h]
110	$C + HCO \leftrightarrow CO + CH$	1.00 (+07)	0.50	9500	-271.85	[h]

schemes for methane combustion [25] and ethene pyrolysis [26] to get the basic set of reactions. The set of reactions involving Ti and free carbon species was pared down by eliminating reactions involving complex internal restructuring and spin forbidden reactions. The final chemical kinetic scheme had 142 reactions as listed in Table III.

Table III - Continued.

Reactions										Forward rate coefficients[a] A (exp)	n	E_a(K)	ΔH_{f-298}	ref
111	CO	+	CH_2	↔	C	+	CH_2O			1.20 (+03)	0.50	39500	328.21	h
112	CH	+	C_2	↔	C_3	+	H			6.65 (+02)	0.50		-397.05	h
113	CH	+	C_2O	↔	C_2	+	HCO			9.86 (+02)	0.50		-4.25	h
114	C_2	+	C_2	↔	C	+	C_3			7.15 (+02)	0.50		-138.73	h
115	CH_2	+	C_2O	↔	C_2	+	CH_2O			1.18 (+03)	0.50		52.11	h
116	TiO	+	M	↔	Ti	+	O	+	M	1.16 (+07)	0.50	79835	668.41	h
117	O	+	TiC	↔	Ti	+	CO			8.39 (+02)	0.79	2500	-542.57	h
118	TiO_2	+	M	↔	TiO	+	O	+	M	1.56 (+07)	0.50	72611	608.99	h
119	CO_2	+	Ti	↔	TiO	+	CO			2.51 (+08)		16911	-136.27	h
120	OH	+	Ti	↔	H	+	TiO			1.48 (+03)	1.18		-240.58	h
121	H	+	TiO_2	↔	OH	+	TiO			4.26 (+05)	0.39	21785	181.16	h
122	TiC	+	M	↔	C	+	Ti	+	M	6.80 (+06)	0.50	64071	533.80	h
123	C	+	TiO	↔	CO	+	Ti			6.93 (+04)	0.50	2500	-407.96	h
124	C	+	TiO_2	↔	CO	+	TiO			1.06 (+05)	0.50	2500	-467.38	h
125	CH	+	Ti	↔	H	+	TiC			1.85 (+04)	0.81		-196.48	h
126	CH	+	TiO	↔	HCO	+	Ti			8.63 (+03)	0.50		-136.11	h
127	CH	+	TiO_2	↔	HCO	+	TiO			1.26 (+03)	0.50		-195.53	h
128	CH_2	+	Ti	↔	H_2	+	TiC			1.25 (+00)	1.16		-207.54	h
129	CH_2	+	TiO	↔	CH_2O	+	Ti			1.03 (+03)	0.50		-79.75	h
130	CH_2	+	TiO_2	↔	CH_2O	+	TiO			1.88 (+02)	0.50		-139.17	h
131	C_2	+	Ti	↔	C	+	TiC			1.20 (+02)	1.48	10000	61.84	h
132	C_2	+	TiO	↔	C_2O	+	Ti			1.08 (+03)	0.50		-131.86	h
133	C_2	+	TiO_2	↔	C_2O	+	TiO			1.50 (+03)	0.50		-191.28	h
134	OH	+	TiC	↔	HCO	+	Ti			1.14 (+02)	1.14	2500	-180.21	h
135	Ti	+	TiO_2	↔	TiO	+	TiO			4.77 (+04)	0.50	1000	-59.42	h

Reaction rates were taken from the literature where possible and estimated for the remaining reactions using transition state theory or collision theory with steric factors as suggested by Moore and Pearson [27, 28]. Some estimates for Ti reactions were made using a comparison with Cr/O compounds [29].

It was assumed that reactions involving free radicals had zero activation energies. While in actual practice this may not be absolutely true, free radical reactions typically have low activation energies. At the higher temperatures in this system, the activation energy term makes only a small change in the reaction rate over a relatively large range of temperatures. Where possible, elementary reactions were written with free radicals as the reactants and with a decrease in energy moving towards products. For reactions written in this manner, the activation energy could be assumed to be small.

In general, at plasma processing temperatures, the relative fraction of collision pairs that have sufficient energy between them to overcome the activation barrier for the reaction to proceed is high. Most reactions are collision limited rather than energy limited. Errors in the estimated activation energies, especially for free radical reactions, have only a small effect over a relatively wide range of temperatures. Because most of the molecules are small, the steric factors are usually not more than several orders of magnitude away from unity.

Table III - Continued.

Reactions							Forward rate coefficients[a] A (exp)	n	E_a(K)	ΔH_{f-298}	ref
136	C_2	+ TiC	↔	C_3	+	Ti	2.09 (+00)	0.41	3000	-200.57	h
137	C_2O	+ Ti	↔	CO	+	TiC	9.67 (+04)	0.50	10000	-214.26	h
138	HO_2	+ Ti	↔	OH	+	TiO	6.35 (+06)	0.50	12000	-392.45	h
139	C_2H	+ M	↔	C_2	+	H	4.67 (+10)		62405	520.20	h
140	CH	+ TiC	↔	C_2H	+	Ti	8.86 (+02)	0.50		-245.70	h
141	C_2H	+ H	↔	C_2	+	H_2	9.23 (+04)	0.50	10250	85.20	h
142	C_2H	+ C	↔	C_3	+	H	4.46 (+04)	0.50		-213.20	h

[a] The forward rate coefficient is given as $k = A\,T^n \exp(-E_a/T)$ where the units of A are mol/m^3 and s, E_a is given in units of K, T is the temperature in K.

[b] These M-body reactions have a dependence on the collision partner which is given as a function of temperature, in this work the functions have been evaluated at their upper limits because the temperature range in the model exceeds these limits.

[c] Data was taken from M-body collisions with argon and modified for collisions with diatomics and polyatomics as M = (2.9 * H_2) + (1.2 * O_2) + (18.5 * H_2O) + (2.1 * CO) + (4.3 * CO_2). Other species not specifically included in this list were treated as CO for diatomics and CO_2 for polyatomics.

[d] In this reaction mechanism, electronically excited states were not treated explicitly. The reaction rate was given for reaction with the excited singlet state of CH_2 and that rate was corrected to account for the probability of finding CH_2 in the excited state rather than treating excited molecules as a separate species.

[e] Data for these reactions had a temperature dependence added to account for the dependence of the collision frequency on temperature because of the wide temperature range of in this work.

[f] Reaction rates were estimated using the dissociation energy as the activation energy and a hard sphere collision rate.

[g] This reaction is spin forbidden with C_2 in the ground state and has a correction added to account for the relative fraction of C_2 molecules in the first excited state.

[h] Reaction rates were estimated using transition state theory or collision theroy with steric factors as suggested by Moore and Pearson [28].

Nucleation and Condensation

Nucleation is not a well-understood process. There are two basic approaches to modeling the nucleation process. The capillarity approximation treats nucleation as a fluctuation process while the kinetic approach treats the formation of clusters in a stepwise manner similar to chemical reactions [30-33]. In this work, nucleation was modeled using the capillarity approximation because kinetic modeling requires a knowledge of the properties of the molecules as they form and this information is presently unavailable for TiC clusters. It is also the simpler of the two approaches. With the capillarity approximation, the predicted qualitative behavior is very similar to that with kinetic models. While it does not yield information about the size distribution of the product particles, it does give an average particle size for the products.

Using the capillarity approximation, the rate of formation of clusters of the critical size required for thermodynamic stability is given as;

$$J_{cl} = 5.4 * 10^{33}\left(\frac{P_v}{T}\right)^2 \frac{(\sigma_\infty \mu_v)^{1/2}}{\rho_c} \exp\left\{\frac{-n^* \ln(S)}{2}\right\} \quad \frac{nuclei}{m^3 s} \tag{1}$$

$$n^* = \frac{4\pi}{3}\rho_c\left(\frac{L}{\mu_v}\right)(r^*)^3 \qquad r^* = \frac{2\sigma_\infty \mu_v}{\rho_c RT \ln(S)} \qquad (2), (3)$$

We followed Wegener [32] and assumed this rate is correct to within a multiplicative factor referred to as the replacement factor, Γ. The range reported in the literature for this correction varies over 23 orders of magnitude. However, for cases similar to ours, the literature suggests that the value of Γ should be in the range of 10^{-3} to 10^{+5} [32,33]. In order to better estimate the value, a series of simulations were run with only TiC and argon as the reactants to study the effect of replacement factor on product particle size. It has been shown in experimental work that TiC and a number of other high temperature ceramics will give particle sizes that are consistently in the order of 0.01 to 0.1 μm when synthesized in thermal plasma reactors and allowed to homogeneously nucleate [1,3,5,6,9]. These simulations showed that a replacement factor of 10^{+5} would give the particle sizes in the correct range to match experimental results and this value was used for throughout the calculations. Figure 1 shows a plot of particle size versus replacement factor for several different TiC/Ar ratios.

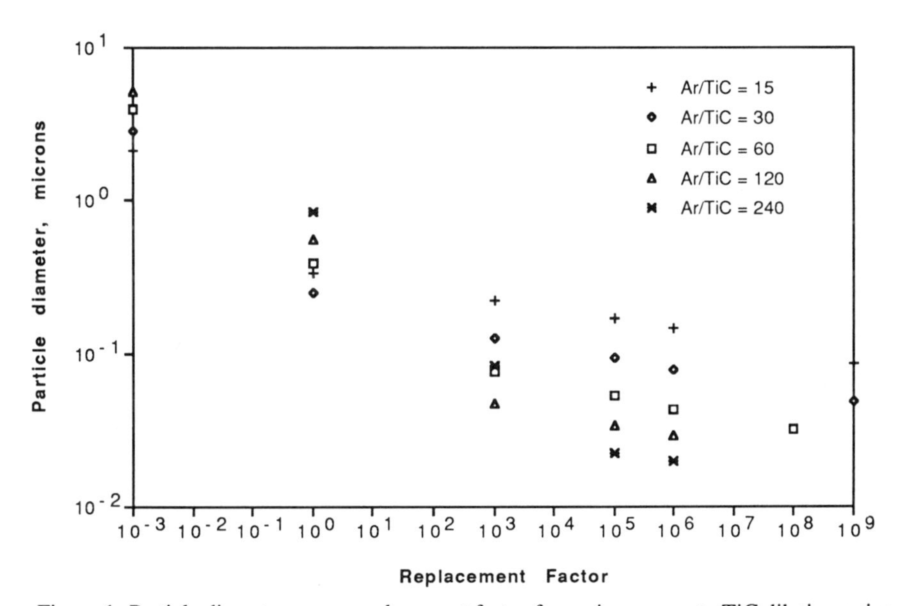

Figure 1- Particle diameter versus replacement factor for various argon to TiC dilution ratios.

The classical rate of nucleation used here does not account for the effect of temperature rise due to the latent heat of vaporization of the condensing species. A correction factor was considered [33]. However, in this case, because of the relatively high mole fraction of hydrogen and the concomitantly high heat transfer rates, the correction factor would be of the order of 1 and was considered unimportant relative to the uncertainty in the replacement factor.

In this model only TiC was considered to homogeneously nucleate since it has the lowest vapor pressure of any of the species in the system and therefore should nucleate first. Nucleation requires supersaturations of the order of 50 to 100 to get significant rates of cluster formation Once there are nuclei present in the system, the other species will tend to heterogeneously condense on the existing nuclei at a rate that is high enough to prevent them from reaching a critical level of supersaturation.

The rate of condensation from the gas phase to the nuclei was determined in several steps. First, the upper limit is the rate of collision between the gas phase molecules and the clusters. Kinetic theory can be used to calculate the collision frequency. In this model, the frequency of collisions is multiplied by a condensation efficiency to give the effective rate of condensation. Each of the condensing species (C, C_2, C_3, Ti, TiC) was assigned its own condensation efficiency. The condensation efficiency was assumed to be a linear function of the ratio of the equilibrium vapor pressure of the species over the droplet to the actual partial pressure of the species in the atmosphere surrounding the droplet and is given by;

$$\eta_{cond} = 1.0 - \frac{P_{equil}}{P_{vapor}} \tag{4}$$

When the partial pressure of the species in the atmosphere surrounding the particle is less than the equilibrium vapor pressure, condensation is assumed not to occur.

The equilibrium vapor pressure over a droplet, i.e. the partial pressure of the species that would be present at equilibrium, is a function of the droplet stoichiometry and the droplet size. With very small droplets, pressure inside the droplet may be much higher than the ambient pressure because of the effect of surface tension and curvature of the droplet. This pressure increases as the droplet size decreases with the consequence that smaller droplets have effectively higher equilibrium vapor pressures over their surfaces than larger droplets. This effect was quantified using the capillarity approximation which assumes that the droplet properties are similar to those of the material in the bulk phase. The surface tension of TiC was estimated to be 1.8 J/m using the techniques described by Richardson [34].

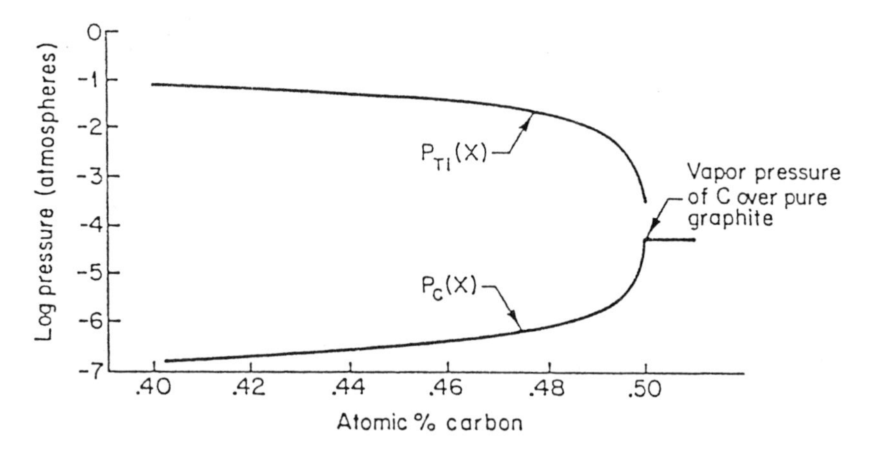

Figure 2 - Calculated vapor pressures over TiC at 3000 K (taken from Toth [35]).

The stoichiometry of the droplet also has a strong effect on the vapor pressure over the droplet. This effect is shown in Figure 2 taken from Toth [35]. In the figure, it is clear that when the stoichiometric ratio of carbon falls to below one, the vapor pressure of carbon over the species falls off rapidly by several orders of magnitude. The effect of stoichiometry on vapor pressure is to affect the relative rates of condensation of carbon and titanium such that the stoichiometry of the particle will tend towards a one-to-one ratio of Ti to C. When the mole fraction of carbon is less then 50 percent, the equilibrium vapor pressure of carbon over the droplet falls giving a subsequent increase in the rate condensation for carbon while at the same time, the vapor pressure of titanium over the droplet increases which tends to slow the rate of condensation of titanium. This effect may explain experimental results [3] where homogeneously nucleated product materials are formed with one-to-one stoichiometric ratios when these ratios are thermodynamically favored only over very narrow ranges.

Condensation rates may also be affected by particle heating due to the latent heat of vaporization of the condensing species. Calculations showed that, again, the relatively high mole fraction of hydrogen in the system played an important role by enhancing heat transfer rates so that the droplet temperatures were maintained very near the ambient temperature during condensation [33].

Soot formation

Soot formation is a complex process that occurs through several different mechanisms depending on the mixture temperature and composition [18,36]. In this work, a relatively simple soot mechanism was modeled in a manner similar to the work of Kiefer et al. [26] where small carbon polymers are assumed to go irreversibly to soot once they are formed. In this case, a C_3 soot mechanism was developed. When C_3 was present at partial pressures above the equilibrium vapor pressure, it was assumed that C_3 participated in a series of reactions to produce C_4 and higher polymers, all of which would then go irreversibly to soot. The C_3 molecule was chosen because the primary composition of the gas phase above graphite is C_3 at the temperatures of interest. When the temperature drops, it becomes thermodynamically favored for C_3 to polymerize to higher carbon polymers or clusters. The reactions considered were C_3 plus other carbon species going to higher polymers. These reactions were assumed to have zero activation energy and the rates were taken as the hard sphere collision probability multiplied by a steric factor [28]. The soot reaction rates were assumed to go in the forward direction only. To take reverse reaction into account when the partial pressure of carbon is near saturation, the reaction rates were multiplied by an efficiency factor. The efficiency was taken as a linear function of the ratio of the equilibrium vapor pressure of C_3 above graphite to the partial pressure of C_3 in the atmosphere and had the same form as the condensation efficiency in equation 4. The reactions and the rates are listed in Table IV.

Table IV - Elementary reactions for irreversible soot formation.

	Reactions	Forward rate coefficients[a]		
		A (exp)	n	E_a(K)
1	$C(g) + C_3(g) \rightarrow C_4(s)$	6.06 (+04)	0.5	0
2	$C_2(g) + C_3(g) \rightarrow C_5(s)$	6.54 (+02)	0.5	0
3	$C_3(g) + C_3(g) \rightarrow C_6(s)$	8.42 (+02)	0.5	0

[a] The forward rate coefficient is given as $k = A\,T^n \exp(-E_a/T)$ where the units of A are mol/m^3 and s, E_a is given in units of K, T is the temperature in K.

Comparison of soot formation versus time for this model with results of a kinetic model by Yoshihara and Ikegami [37] showed that the onset and initial rate of soot formation compared well with their model. However after the onset of soot formation, this model does not describe soot formation well at lower temperatures.

Most soot models rely on larger polyacetylenic species and more complex hydrocarbons which we do not expect to find in this system. At higher temperatures, soot formation is thought to occur by the carbon molecule association rather than via more complex molecules [38]. While this model is somewhat simplistic, it does incorporate a number of features that should give reasonable qualitative behavior. It has a dependence on partial pressure both for soot initiation and for control of the rate at which carbon is lost to soot. It is also limited by kinetics for the formation of free carbon to be available to go to soot. This latter mechanism ties soot formation into the chemical kinetics of the system.

We have not included the condensation of titanium directly onto soot particles. While this may occur, experiments show that metal carbides often form as a homogeneously nucleated particle covered by turbostratic soot particles. This is direct evidence that soot formation and homogenous nucleation occur in parallel as competing mechanisms [39].

Numerical Implementation

The chemical kinetics, nucleation and growth, and soot mechanisms were incorporated into a numerical model that was run at fixed pressures with temperature profiles based on the plasma torch our group has been using for experimental work. The variables included all the reactive species, number of particles, moles of carbon and titanium nucleated or condensed to the particles, the number of moles of soot per kg of mixture and temperature. The basic temperature versus time profile is a rapid linear rise in temperature and several ms residence time at the peak temperature followed by an exponential decrease in temperature as shown in Figure 3. The initial rate of cooling is approximately $6 * 10^5$ K/s. The models were run for 50 ms at which point the system temperature had fallen to around 1000 K and no further reaction was considered.

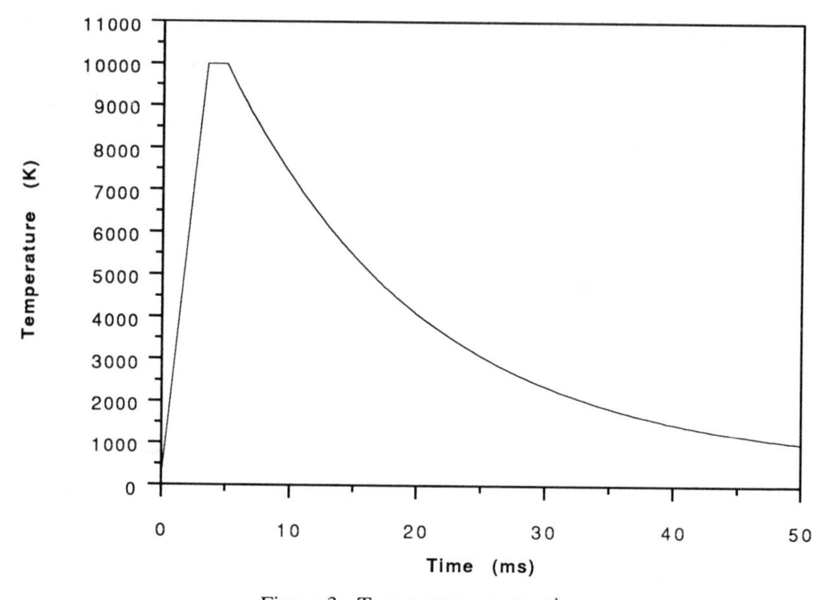

Figure 3 - Temperature versus time.

The model was integrated through time using the STRIDE package which is a singly implicit Runge-Kutta technique specifically for large systems of stiff differential equations [40]. The method has automatic step size, order and method control and has an advantage over multi-step backwards differentiation methods in that it is very stable at high orders. Because of the stiffness of the set of governing equations, the integration required as high as tenth order terms during the integration.

The program was approximately 7,500 lines of FORTRAN source code. The simulations were run on a cluster of IBM RS6000 series workstations with each run typically taking 10 to 60 minutes of CPU time.

Results and Discussion

A large number of simulations were run over a wide range of conditions. General trends are shown Figures 4 through 7. The parameters varied included CH_4 to TiO_2 ratio, degree of inert gas dilution, the addition of hydrogen and system pressure. The data plotted includes the supersaturation of TiC at the point the partial pressure of C_3 reached saturation (S-Ti), the Ti to C ratio in the product particle (Ti/C), the percentage of condensed phase carbon appearing as soot (% Soot) and the sum of the concentrations of CH_4, C_2H_2 and C_2H_4 (C_yH_x mol/m^3). The supersaturation of TiC when the partial pressure of C_3 reaches saturation is a measure of the propensity of the system to form soot. When the supersaturation ratio is high enough to drive nucleation in advance of soot formation, there will be less soot in the product since carbon leaves the gas phase through nucleation and subsequent growth of TiC. The final ratio of Ti to C in the product particles is a measure of how much carbon was available for product formation without being lost to soot of leaving the system in gas phase species. Also shown in Figures 8 through 10 are plots of nucleation, sooting and chemical concentration data for a typical simulation.

Figure 4 shows data for simulations run with a stoichiometric ratio of CH_4 to TiO_2 (3 to 1) at different argon dilution rates with no H_2 addition. The maximum supersaturation of TiC is about 15 and in all cases there is a high proportion of soot in the product because the rate of nucleation is still insignificant when sooting starts to occur. The product particles are predicted to have Ti to C ratios in the range of 3 to 5 which is a substantially substoichiometric composition. It should be noted that although the performance appears to improve as the degree of inert gas dilution is decreased, this would become increasingly more difficult in practice. Typical inert gas dilution ratios are in the order of 10 to 100 [16].

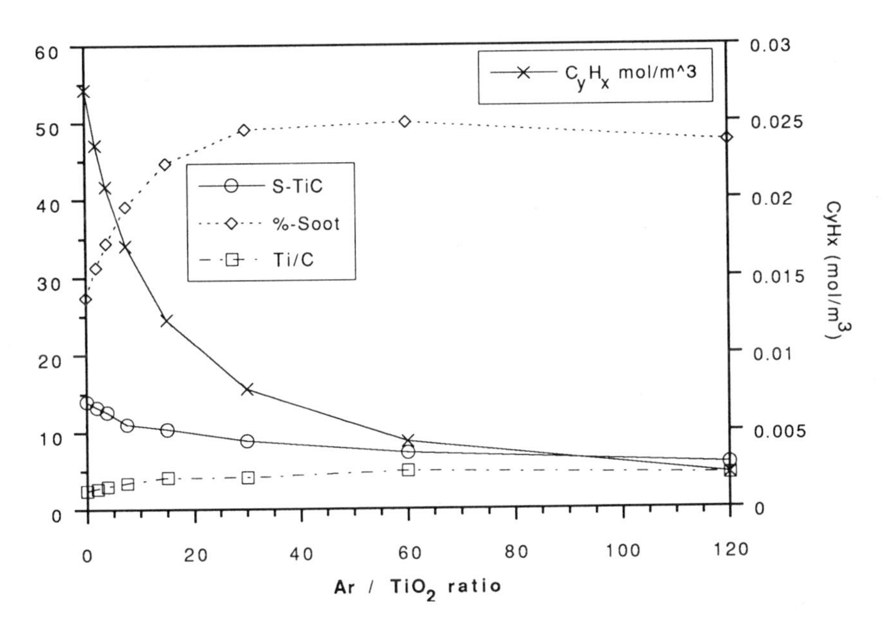

Figure 4 - The effect of varying argon dilution with CH_4 / TiO_2 = 3.0, H_2 / TiO_2 = 0.0, P = 1.0 atm.

Figure 5 shows data for a fixed argon to TiO_2 dilution of 60 to 1 and a fixed CH_4 to TiO_2 ratio of 6 to 1. The mixture is carbon rich to try to get better Ti to C ratios in the product. Varying amounts of hydrogen were added to the system to try to inhibit soot formation. Clearly, the addition of hydrogen is helpful. However, much of the carbon becomes bound with hydrogen so that as the fraction of H_2 increases and soot decreases, not enough carbon is available to form TiC and the supersaturation is still low (~ 1 - 10) when C_3 becomes saturated. The result is that there is a high fraction of soot formed and the stoichiometric ratio of the product is in range of 2 to 10 Ti/C. Also shown in this plot is the sum of the concentrations of CH_4, C_2H_2 and C_2H_4 (C_yH_x). With no additional hydrogen added to the system about 5 percent of the total carbon leaves the system as a gaseous product. At the highest dilution of 120 to 1 about 73 percent of the total carbon leaves the system as a gaseous product. The decrease in concentration of C_yH_x is due to the dilution by hydrogen.

Figure 6 shows the dependence of the system on pressure for a fixed argon dilution ratio of 60 to 1, a CH_4 to TiO_2 ratio of 6 to 1 and a H_2 ratio of 60 to 1. The pressure is varied between 1 and 16 atm. The supersaturation of TiC at C_3 saturation (the onset of sooting) increases steeply with pressure. After the pressure reaches 8 atm, the system never becomes saturated in C_3 because nucleation and growth of TiC remove so much carbon from the system. While the results improve with pressure, the ratio of Ti to C only falls to a little over 2 and levels off because there is an appreciable amount of carbon that is unavailable for TiC formation. The fraction of carbon in the gas phase products is about 26 percent and varies little with pressure.

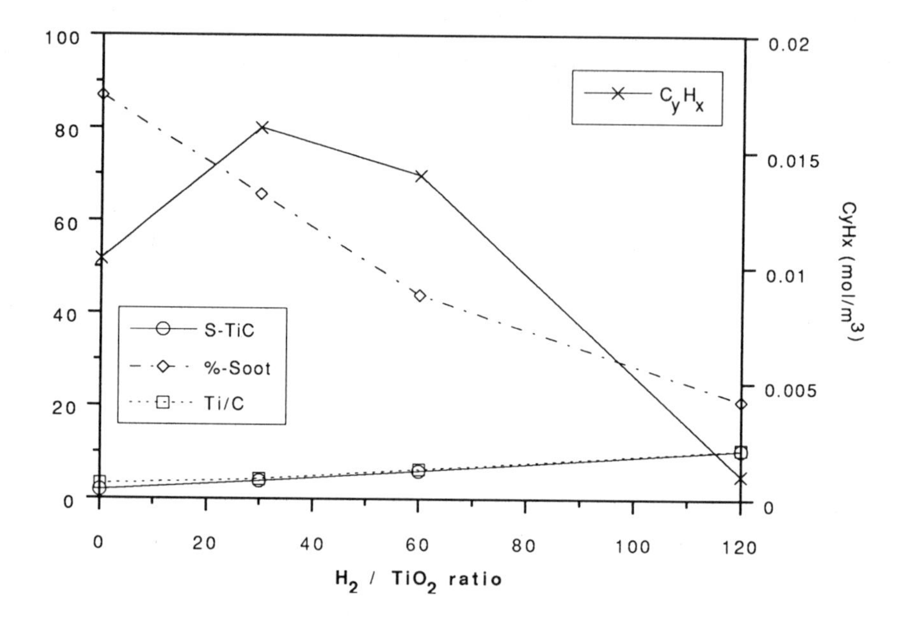

Figure 5 - The effect of hydrogen addition with
CH_4 / TiO_2 = 6.0, Ar / TiO_2 = 60.0, P = 1.0 atm.

Figure 7 shows data for conditions similar to those in Figures 5 and 6. The argon dilution ratio is 60 to 1, the H_2 addition ratio is 60 to 1 and the pressure is now fixed at 16 atm. The CH_4 to TiO_2 ratio is varied to try to find conditions for synthesizing a stoichiometric product. It can be seen that the stoichiometric ratio of Ti to C approaches one as the CH_4 addition ratio approaches 7.9. At this point there is a sharp increase in the amount of soot produced corresponding to the steep decrease in TiC supersaturation. This is caused by carbon forming more C_3 relative to TiC thus raising the partial pressure of C_3 and reducing the partial pressure of TiC. When the mixture is leaner at a CH_4 of 7.5, a product is formed that is just slightly substoichiometric with a soot fraction of about 10 percent. Figures 8 through 10 show data for nucleation, soot formation and chemical concentration for the case with a CH_4 ratio = 7.9.

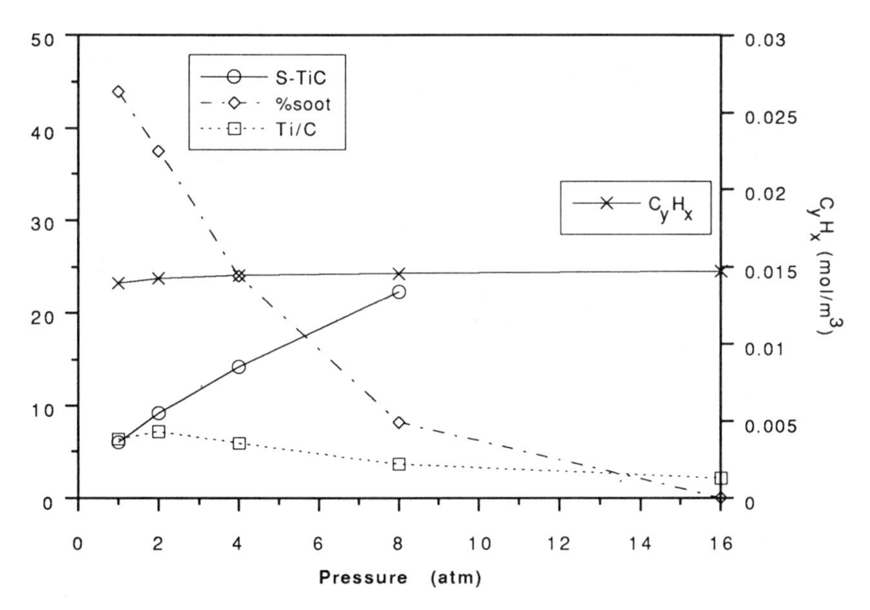

Figure 6 - The effect of pressure with
CH_4 / TiO_2 = 6.0, Ar / TiO_2 = 60.0, H_2 / TiO_2 = 60.0.

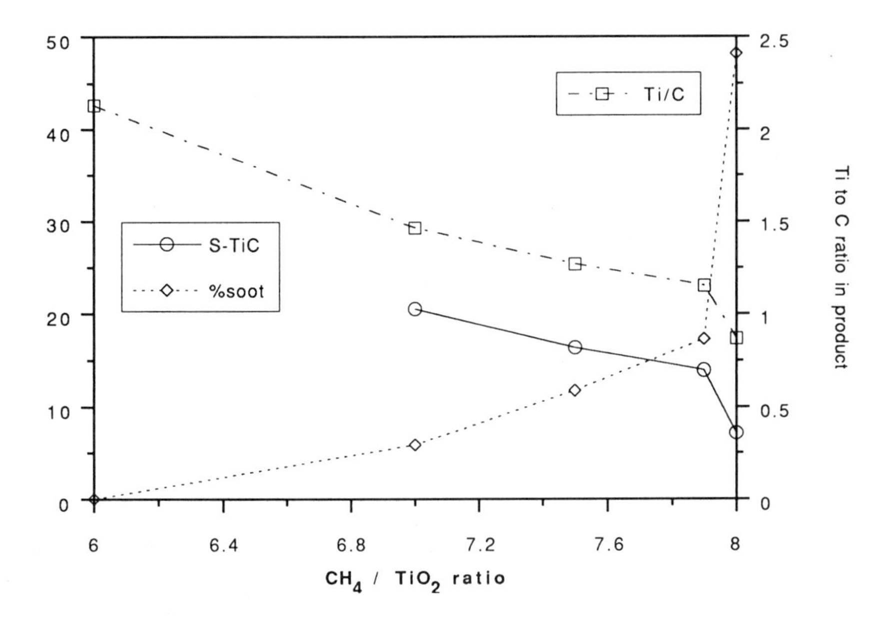

Figure 7 - The effect of CH_4 ratio with
Ar / TiO_2 = 60.0, H_2 / TiO_2 = 60.0, P = 16 atm

In Figure 8, a plot of nucleation rate versus temperature shows the extreme dependence of nucleation rate on temperature. The nucleation rate (J_{cl}) rises from virtually zero to about 10^{18} over a range of approximately 150 K which corresponds to a period of ~ 1 ms with the given rate of temperature change. This rate is also dependent on the number of TiC molecules in the critical stable cluster size (n^*) which is also shown on the plot. The number drops from ~ 200 to ~ 20 as the supersaturation of TiC rises from ~ 1 to ~ 15. Also shown on this plot are the total number of nuclei (#nuc). The nuclei form quickly and as this happens, the partial pressure of TiC falls off, increasing critical cluster size and giving a sharp decrease in nucleation rate. Once the initial burst of nucleation is over, the clusters grow by heterogeneous condensation and the partial pressure of TiC stays below the critical point. Figure 9 shows soot formation rate versus temperature for the same conditions. Both phenomena occur virtually at the same time, although by different mechanisms (which may explain why they are often observed to occur in parallel). The soot rate (dS/dt) is given in units of moles of carbon per kg of mixture going to soot per second. It rises steeply as C_3 reaches saturation (S-C_3) and the bulk of the soot is formed over a period of ~ 0.5 ms. The rapid falloff in soot rate is due to both the loss of carbon to soot as well as the loss of carbon through nucleation and condensation to TiC.

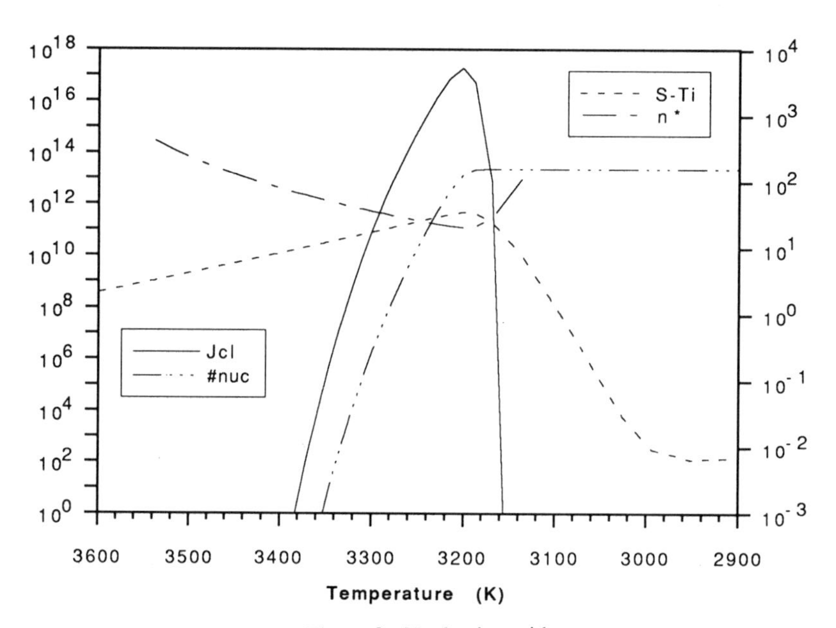

Figure 8 - Nucleation with
$CH_4 / TiO_2 = 7.9$, $Ar / TiO_2 = 60.0$, $H_2 / TiO_2 = 60.0$, P = 16 atm.

Figure 10 shows the concentrations of some of the species in the system along with soot concentration. The formation of C_2H_2 is seen to begin well before either soot formation or nucleation in the system. On this plot, nucleation and soot formation both occur at the point on the plot where soot concentration is nearly a vertical line at around 3400 K. Once soot formation and TiC particle formation and growth begin, the relative amount of available carbon in the system can be seen to drop as C_3 falls off sharply. At this point, the rate of formation of C_2H_2 levels off as there is no more carbon accessible for reaction. The C_2H_2 level begins to fall when the temperature drops further to around 2250 K when the reaction to C_2H_4 becomes favored.

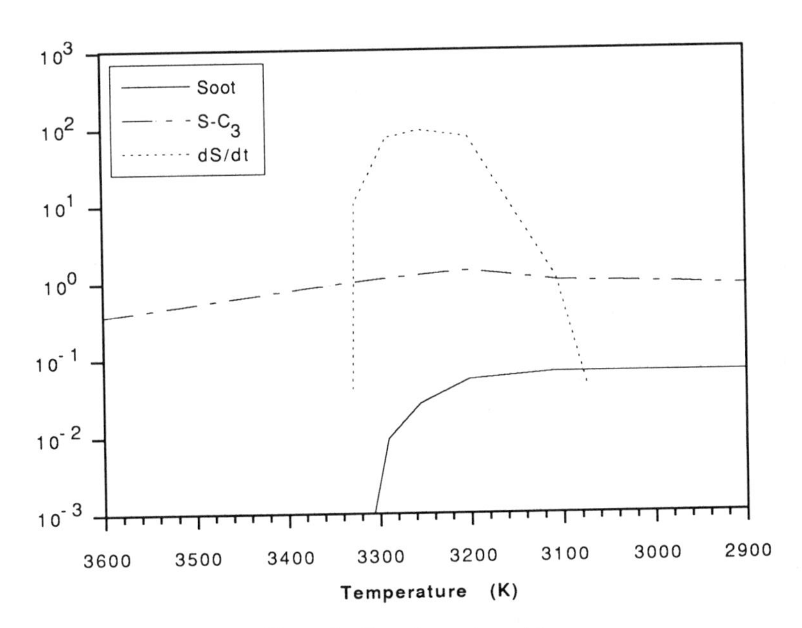

Figure 9 - Soot formation with
$CH_4 / TiO_2 = 7.9$, $Ar / TiO_2 = 60.0$, $H_2 / TiO_2 = 60.0$, P = 16 atm.

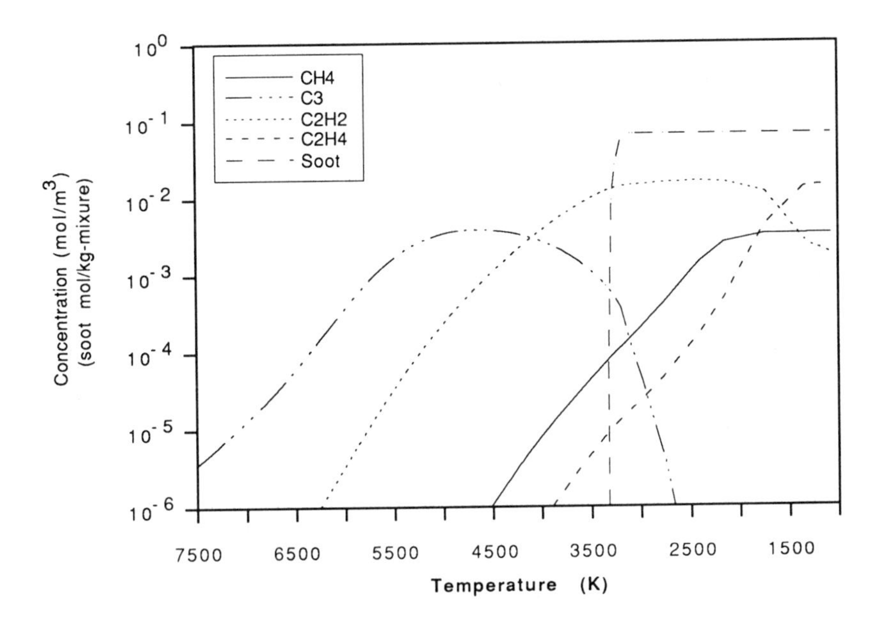

Figure 10 - Chemical concentration with
$CH_4 / TiO_2 = 7.9$, $Ar / TiO_2 = 60.0$, $H_2 / TiO_2 = 60.0$, P = 16 atm.

The results show that there is a strong tendency for the system to form substoichiometric TiC plus soot over a wide range of conditions. The formation of soot agrees with observations from our experiments and with reported experiments using other transition metal carbides [3,6-10]. Where soot has not been reported and X-ray diffraction (XRD) has been used for product analysis, it is suspected that there may well have been soot present. As soot, carbon would be in an amorphous form and only show up as background noise in XRD.

There is a tradeoff between the competing processes of soot formation and the nucleation and growth of titanium carbide particles. Essentially, if titanium carbide nucleates first, then it is preferable for Ti, C and TiC molecules to condense onto the existing TiC nuclei. When this happens, the rate of condensation of these species keeps the partial pressures low enough so that homogeneous nucleation is never favored for the remaining species. The formation of soot, however, is a kinetic process and requires only that the atmosphere be saturated in carbon before irreversible soot formation begins to occur. If soot formation occurs before the nucleation of TiC, then carbon very quickly leaves the atmosphere and the partial pressure of carbon rapidly falls to the equilibrium partial pressure for carbon and remains there. As the temperature falls further and TiC nucleates, particle growth is then primarily by Ti condensation as much of the available carbon is lost to soot. The resulting particles are predicted to have very high Ti to C ratios. High mole fractions of hydrogen may be used to try to inhibit the formation of soot by combining with the carbon and keeping its vapor pressure relatively low while still leaving the carbon in a form that is chemically available for the formation of TiC.

The system was shown to respond to increasing pressure and to increasing ratios of hydrogen to carbon. However, both of these effects also increased the fraction of acetylene and ethene in the products and increased the carbon requirement to form products with an acceptable Ti to C ratio. Even at the higher pressures, it would be difficult to get a stoichiometric product without an appreciable amount of soot.

In practice, there are limits to the amount of hydrogen that can be added to the gas because of the behavior of the plasma torch. Typical experimental RF plasma torches have relatively low power and the addition of a high proportion of diatomic gases to the plasma will often extinguish the torch because dissociation absorbs to much power. Typically, a minimum of 50 kW would be required to sustain an RF hydrogen plasma. Hydrogen will also to the heat load on the torch wall because of the enhanced heat transfer.

Conclusion

The model results compare well with experiential results. The key feature that has been highlighted is that the competition between the nucleation and growth of the product and the formation of soot will be the controlling parameter in the thermal plasma processing of TiC powders. While TiC has a much lower vapor pressure than carbon, homogeneous nucleation requires significant supersaturations, while soot formation is a kinetic process requiring only that the system reach carbon saturation before sooting begins. In order to produce TiC powders that are free from soot, the partial pressure of TiC must be raised while the partial pressure of carbon is decreased or held constant. This can be accomplished by adding hydrogen to the system. However, the addition of hydrogen alone is not sufficient to give good product stoichiometries at atmospheric pressure. At higher pressures, the partial pressure of TiC is raised relative to carbon. Pressure may provide means for achieving acceptable process performance, although the process still requires a high mole fraction of hydrogen and the balance between soot formation and Ti to C ratio is delicate.

The current model we are using for soot formation is limited and provides qualitative information only, although it is expected that the prediction for the onset of sooting is relatively accurate. In the model, the rate of soot formation is kinetically controlled and tied to the ratio of the equilibrium vapor pressure and the actual partial pressure of C3 in the system. However the kinetic mechanism is probably too limited to give an accurate representation, especially at lower temperatures where polyacetylenic species would be expected to form. Current work is underway to implement a more detailed kinetic mechanism for soot formation into the model so that carbon must go through a series of reversible kinetic steps before being lost to soot.

With more information about TiC clusters, a kinetic model for homogeneous nucleation could be implemented into the model. It would be expected that small TiC clusters may alter the chemistry of the system by increasing the amount of carbon tied up with titanium in the gas phase. This could lower the partial pressure of carbon thereby delaying soot formation and also increase the rate of condensation of TiC to the nuclei once they have formed. TiC clusters with unusual stoichiometries such as Ti_8C_{12} have recently been reported [58]. A kinetic model would also give some indication as to expected the particle size distributions and possibly an estimate of agglomeration as well.

The uncertainty in the properties of gas phase TiC is also likely to significantly affect model behavior. Currently our group is undertaking high accuracy *ab initio* quantum chemical calculations on the properties of gas phase titanium based ceramic molecules. The results of this study should prove useful for further work in this area, especially since this data will affect the vapor pressure of TiC and the size of the processing window. It would also be appropriate to perform a sensitivity analysis on the reaction mechanism to identify the more important reactions for more refined rate estimates.

To get a good product quality without soot formation in an RF torch at atmospheric pressure, nucleation and growth will have to be enhanced either by the addition of seed nuclei or by the chemical means. Seed could be added after the plasma to provide a heterogeneous condensation site while the supersaturation is still too low to drive a homogeneous nucleation. It is possible that the addition of Hf or Ta might help to from nuclei well in advance of soot formation. Further study will have to be done to model the homogeneous nucleation of Ti, Hf, Ta and C mixtures. There is also the possibility that reactive quenching may be a way to introduce carbon into the system at the appropriate time to avoid soot formation.

Acknowledgments

This program was supported in part by,

- The New Zealand Department of Scientific and Industrial Research, Chemistry, Contract No. UV/CD/3/1.

- The New Zealand Lottery Science Board.

- The Auckland University Research Council.

- IBM New Zealand.

Symbols

B_e	rotational constant
D_o	dissociation energy
g_i	electronic degeneracy
ΔH_f^o	enthalpy of formation
$\Delta H_v(T_b)$	enthalpy of vaporization at boiling point
J_{cl}	rate of cluster formation, #/m^3-s
L	Avagadro's number
n^*	critical stable cluster number
P_{equil}	equilibrium vapor pressure
P_{vapor}	partial pressure of a species
r^*	critical stable cluster radius
R	ideal gas constant
S	supersaturation
ε	electronic energy level
η_{cond}	sticking efficiency of molecules during condensation
μ_v	molecular weight of gas phase molecules
ρ_c	density of condensed phase
σ	symmetry number
σ_∞	surface tension
ω_e	vibrational constant

References

1. S.R. Blackburn, T.A. Egerton and A.G. Jones, "Vapour Phase Synthesis of Nitride Ceramic Powders Using a D.C. Plasma," Paper 47, Fine Ceramic Powder Conference, Warick, April (1990)

2. A.G. Jones, "Tioxide Titanium Nitride, " CB 1.1, Tioxide Chemicals Ltd., (1991)

3. P.C. Kong, M. Suzuki, R. Young and E. Pfender, "Synthesis of Beta-WC_{1-x} in an Atmospheric-Pressure, Thermal Plasma Jet Reactor," Plamsa Chem. Plasma Process., **3**, 1, 115-133 (1983)

4. R. McPherson, "Plasma Processing of Ceramics," J. Aust. Ceram. Soc., **17**, 1, 2-5 (1981)

5. E. Neuenschwander, "Herstellung und Charakterisierung von Ultrafeinen Karbiden, Nitriden und Metallen," J. Less-Common Met., **11**, 365-375 (1966)

6. B. Mitrofanov, A. Mazza, E. Pfender, P. Ronsheim and L.E. Toth, "D.C. Arc Plasma Titanium and Vanadium Compound Synthesis from Metal Powders and Gas Phase Non-Metals," Mat. Sci. Eng., **48**, 21-26 (1981)

7. D.S. Phillips and G.J. Vogt, "Plasma Synthesis of Ceramic Powders," MRS Bulletin, October 1/November 15, 54-58 (1987)

8. T. Yoshida, "Thermal Plasma Synthesis of Ceramic Powders and Coatings," in Z.A. Munir and J.B. Holt (eds.), "Combustion and Plasma Synthesis of High-Temperature Materials," VCH Publ., New York, 328-339 (1990)

9. Z.P. Lu, T.W. Or, L. Stachowicz, P. Kong and E. Pfender, "Synthesis of Zirconium Carbide in a Triple Torch Plasma Reactor Using Liquid Organometallic Zirconium Precursors," Mat. Res. Soc. Symp. Proc., **190**, 77-82 (1990)

10. P.R. Taylor and S.A. Pirzada, "Synthesis of Ultrafine Silicon Carbide in a Non-Transferred Arc Thermal Plasma Reactor," in K. Upadhya(ed), "Plasma and Laser Processing of Materials," TMS, 123-139 (1991)

11. S.V. Joshi, Q. Liang, J.Y. Park and J.A. Batdorf, "Effect of Quenching Conditions on Particle Formation and Growth in Thermal Plasma Synthesis of Fine Powders," Plasma Chem. Plasma Proc., **10**, 2, 339-358 (1990)

12. J.S. McFeaters and J.J. Moore, "Application of Nonequilibrium Gas-Dynamic Techniques to the Plasma Synthsis of Ceramic Powders," in Z.A. Munir and J.B. Holt (eds.), "Combustion and Plasma Synthesis of High-Temperature Materials," VCH Publ., New York, 431-446 (1990)

13. F.H.A.G. Fey, W.W. Stoffels, E. Stoffels, J.A.M. van der Mullen, B. van der Sijde and D.C. Schram, "Kinetics of an Argon Inductively Coupled Plasma," in U. Ehelmann, H.G. Lergon and K. Wiesemann (eds.), "10th International Symposium on Plasma Chemistry," Bochum, Aug. 4-9, 1.1-18 p.1-6 (1991)

14. M.H. Gordon and C.H. Kruger, "Electronic Quenching of Argon Excited States in a Non-Equilibrium Plasma at Atmospheric Pressure," in U. Ehelmann, H.G. Lergon and K. Wiesemann (eds.), "10th International Symposium on Plasma Chemistry," Bochum, Aug. 4-9, 1.2-9 p.1-6 (1991)

15. Y. Chang and E. Pfender, "Thermochemistry of Thermal Plasma Chemical Reactions. Part I. General Rules for the Prediction of Products', Plasma Chem. Plasma Proc., **7**, 3, 275-297 (1987)

16. R.L. Stephens, M.K. Wu, B.J. Welch, J.S. McFeaters and J.J. Moore, "A Thermodynamic Analysis of Titanium Carbide Synthesis in a Thermal Plasma Reactor," in K. Upadhya (ed.), "Plasma Synthesis and Processing of Materials," TMS (1993)

17. M.Z. Luo, J.S. McFeaters and G.D. Mallinson, "Numerical Modelling of an RF Plasma Torch with Swirl," 11th Australasian Fluid Mechanics Conference, Hobart, Tasmania, Dec. 14-18 (1992)

18. H.F. Calcote, "Mechanisms of Soot Nucleation in Flames - A Critical Review," Combust. Flame, **42**, 215-242 (1981)

19. S.E. Stein and A. Fahr, "High-Temperature Stabilities of Hydrocarbons," J. Phys. Chem., **89**, 17, 3174-3725 (1985)

20. M.W. Chase, C.A. Davies, J.R. Downey, D.J. Frurip, R.A. McDonald and A.N. Syverud, "JANAF Thermochemical Tables," 3rd Ed., J. Phys. Chem. Ref. Data, Supplement No. 1, **14** (1985)

21. C.W. Bauschlicher Jr. and E.M. Siegbahn, "On the Low-Lying States of TiC," Chem. Phys. Lett., **104**, 4, 331-335 (1984)

22. A.G. Gaydon, "Dissociation Energies and Spectra of Diatomic Molecules," Academic Press, New Yory, p285 (1968)

23. J.S. McFeaters, "The Non-Equilibrium Gas Dynamic Synthesis Of Transition Metal Carbide Powders," PhD Thesis, Carnegie-Mellon University (1986)

24. G. Herzberg, "Molecular Spectra and Molecular Structure: IV. Constants of Diatomic Molecules," 2nd ed., Van Nostrad, New York (1979)

25. M. Frenklach, H. Wang and M.J. Rabinowitz, "Optimization and Analysis of Large Chemical Kinetic Mechanisms Using the Solution Mapping Method - Combustion of Methane," Prog. Energy Combust. Sci., 18 (1992), 47-73.

26. J.H. Kiefer, S.A. Kapsalis, M.Z. Al-Alami and K.A. Budach, "The very High Termpature Pyrolysis of Ethylene and the Subsequent Reactions of the Product Acetylene," Combust. and Flame, 51 (1983), 79-93.

27. S.W. Benson,"Thermochemical Kinetics, " (New york NY:John Wiley and sons, 1976).

28. J.W. Moore and R.G. Pearson, "Kinetics and Mechanism," (New York NY:Wiley, 1981).

29. U.S. Akhmadov, I.S. Zaslonko, V.N. Smirnov, "Mehcanisms and Kinetics of the Interaction of Fe, Cr, Mo and Mn," Kint. Catal., 29 (1988), 251.

30. S.L. Girshick and C.-P. Chiu, "Homogenous Nucleation of Particles from the Vapor Phase in Thermal Plasma Synthesis,"Plasma Chem. Plasma Proc., **9**, 3, 355-369 (1989)

31. A.W. Castleman Jr., "Clusters: Elucidating Gas-to-Particle Conversion Processes," Environ. Sci. Technol., **22**, 11, 1265-1267 (1988)

32. P.P. Wegener, Nonequilibrium Flows, Part I, (New york NY: Marcel Dekker, (1969).

33. A.C. Zettlemoyer, ed., Nucleation Phenomena, (Elsevier, 1977), 369, 373.

34. F.D. Richardson, Physical Chemistry of Melts in Metallurgy, (Pergamon, 1979), 426.

35. L.E. Toth, Transition Metal Carbides and Nitrides, Vol. 7 in "Refractory Materials,"J.L. Margraves (ed.), Academic Press, New York, (1971), 136.

36. M. Frenklach, D.W. Clary, T. Yuan, W.E. Gardiner Jr. and S.E. Stein, "Mechanism of Soot Formation in Acetylene-Oxygen Mixtures," Combust. Sci. Tech., **50** (1986), 79-115.

37. Y. Yoshihara and M. Ikegami, "Homogeneous Nucleation Theory for Soot Formation," JSME Int. J., Series II, **32**, 2 (1989), 273-280.

38. K. Shakourzadeh Bolouri and J. Amouroux, "Reactor Design and Energy Concepts for a Plasma Process of Acetylene Black Production," Plasma Chem. Plasma Process., **6**, 4, (1986) 335-348.

39. K. Kinoshita, "Carbon: Electrochemical and Physicochemical Properties," (New York: John Wiley and Sons, 1988), 13, 22, 42.

40. K. Burrage, J.C. Butcher and F.H. Chipman, "An Implementation of Singly-Implicit Runge-Kutta Methods," BIT, 20 (1980), 326-340.

41. W. Tsang and R.F. Hampson, "Chemical Kinetic Data Base for Combustion Chemistry. Part 1. Methane and Related Compounds," J. Phys. Chem. Ref. Data, vol 15, 3 (1986), 1087-1279.

42. J.N. Murrel and J.A. Rodriguez, "Predicted Rates Constants for the Exothermic Reactions of ground State Oxygen Atoms and CH Radicals," J. Molec. Struct. (Theochem), 139 (1986), 267.

43. J. Warnatz, "Rate coefficients in the C/H/O system, " Combustion Chemistry, (New York NY: Springer Verlag, (1984)

44. J. Vandooren, O. de Guertechin and P. van Tiggelen, "Kinetics in a Lean Formaldehyde Flame," Combust. Flame, 64 (1986), 127.

45. A.M. Dean and P.R. Westmoreland, "Bimolecular QRRK Analysis of Methyl Radical Reactions," Int. J. Chem. Kinet., 19 (1987), 207.

46. P. Frank and Th. Just, "High Temperature Kinetics of Ethylene-Oxygen Reaction," Symp. Int. Shock Tube Proc., 14 (1984), 706.

47. M.R. Berman and M.C. Lin, "Kinetics and Mechanism of the Reactions of CH with CH_4, C_2H_6 and n-C_4H_{10}," Chem. Phys., 82 (1983), 435.

48. M.R. Berman, J. Fleming, A. Harvey and M. Lin, "Temperature Dependence of the CH Radical Reactions with O_2, NO, CO and CO_2," Proc. Int. Combust., 19 (1982), 73.

49. A.J. Dean, D.F. Davidson and R.K. Hanson, "A Shock Tube Study of C atoms with H_2 and O_2 using Excimer Photolysis of C_3O_2 and C Atom Atomic Resonance Absorption Spectroscopy," J. Phys. Chem., 95 (1991), 183.

50. M.W. Slack, "Kinetics and Thermodynamics of the CN Molecule III," J. Chem. Phys., 64 (1976), 228.

51. P. Frank, "A high Temperature Shock Tube Study on Fast Reactions of Methylene and Methyl Radicals," Proc. Int. Symp. Rarefied Gas Dyn., 2 (1986), 422.

52. D. Husain and L.J. Kirsch, "Reactions of Atomic Carbon by Kinetic Absorption Spectroscopy in the Vacuum Ultra-Violet," Trans. Faraday Soc., 67 (1971), 2025.

53. W. Bauer, K.H. Becker and R. Meuser, "Laser Induced Fluorescence Studies on C_2O and CH Radicals," Ber. Binsenges. Phys. Chem., 89 (1985), 340.

54. V.M. Donnely, W.M. Pitts and J.R. McDonald, "$C_2O(X^{3-})$: Absolute Reaction Rates Measured by Laser Induced Fluorescence," Chem. Phys., 49 (1980), 289.

55. D. Husain and A. N. Young, "Kinetic Investigation of Ground State Carbon Atoms," J. Chem. Soc. Faraday Trans., 71 (1975), 525.

56. H. Reisler, M. Mangir and C. Wittig, "Kinetics of Free Radicals Generated by IR Laser Photolysis. IV. Intersystem Crossings and Reactions of $C_2(X_1g^+)$ and $C_2(A_3u)$ in the Gaseous Phase," J. Chem. Phys., 73 (1980), 2280.

57. J.S. McFeaters, "The Non-Equilibrium Gas Dynamic Synthesis of Transition Metal Carbide Powders," PhD. Thesis, Carnegie Mellon University, (1986).

58. B.C. Guo, K.P. Kearns and A.W. Castleman, "$Ti_8C_{12}^+$ - Metalllo-Carbohedrenes: A New Class of Molecular Clusters?," Science, 255 (1992), 1411.

GENERATION OF NANO-CRYSTALLINE METALS IN A TRANSFERRED ARC THERMAL PLASMA REACTOR

P.R. Taylor, S.A. Pirzada, D.L. Marshall, and S.M. Donahue

Dept. of Metallurgical Engineering, College of Mines
University of Idaho, Moscow, Idaho 83843

ABSTRACT

Plasma processing has the potential to generate nanometer sized metallic powders, which have enhanced physical and mechanical properties. A novel transferred arc thermal plasma reactor has been designed, constructed and operated to perform this synthesis. The reactor provides an extended flame zone which ensures complete vaporization of the powdered metallic feed material. This complete vaporization is a prerequisite to the generation of nano-sized materials by rapid quenching. Experimental results are presented on the generation of nanometer sized aluminum, nickel and iron powders.

Plasma Synthesis and Processing of Materials
Edited by Kamleshwar Upadhya

Introduction

Nanocrystalline metals or materials are solids with a grain size of a few nanometers (1-50 nm)[1]. Nanosize materials can be metals, ceramics or composites. Due to their extremely small size, these materials are comprised of two structural components: [2] crystallites with long range order and the disordered interfacial component representing a variety of atomic spacings in the different types of interfaces. Due to a substantial disordered interfacial component, these nanosize materials show enhanced physical and mechanical properties. These properties include high diffusivity, reduced density, increased thermal expansion and specific heat, alloying of conventionally insoluble or low solubility elements and the ductile behavior of the nanocrystalline ceramics and intermetallic compounds [3]. Several investigators have investigated the structural aspects of nanocrystalline material and have proposed several mechanisms which seem to be responsible for the several enhanced properties of these materials [4-8]. The state of information on this new class of materials is in the early stage of development.

There are several ways to generate or synthesize nanocrystalline materials. Gleiter explained a gas-condensation method for producing nanosized materials [4]. The system consists mainly of an ultrahigh vacuum (UHV) chamber which has two resistively heated evaporation sources, a cluster collection device (cold finger) and scraper assembly. During evaporation in an inert atmosphere (He), convective gas flow transports the particles, which are formed by condensation in the region close to the source, to a liquid nitrogen filled cold finger, where the particles are collected and the nanophase compacts are formed in a compaction device. Layered nanostructures are produced by vapor deposition or by electrodeposition processes. There are other potential methods for preparing nanosize materials, such as mechanical alloying, molecular beam epitaxy, rapid solidification, ion beam, reactive sputtering, sol-gel and chemical vapor deposition.[3]

Plasma synthesis of pure and fine (submicron) ceramic and metallic powders is a widely studied area of plasma processing. Numerous laboratory systems have been designed and operated to generate fine (submicron) ceramic materials (such as carbides, nitrides, oxides, etc.)[10-12]. Plasma processing may also have potential for the generation of nanocrystalline metals/materials. There have been some successful attempts to produce nanosize materials in plasma reactors [13-14]. An important attraction in using plasma reactors for the generation of nanocrystalline materials seems to be the possibility of producing substantial quantities of these types of materials in a very clean atmosphere. The highly concentrated enthalpy available in the plasma flame has the ability to vaporize virtually any refractory material provided the reactant is properly injected into the plasma arc and has sufficient residence time in the hot zone. The steep temperature gradients available in plasma reactors may lead to homogeneous nucleation which can result in the generation of nanosize materials.

This work reviews the development of a transferred arc thermal plasma reactor system to generate nanocrystalline metals.

A novel transferred arc plasma system has been designed, built and operated. Preliminary experiments are performed to generate nanocrystalline sized powder of metals such as aluminum, nickel and iron. Generating nanosize metal powders is the first stage of this work, while the next step will be to generate nanocrystalline intermetallics. Intermetallics such as titanium aluminide (TiAl, Ti_3Al) and nickel aluminide (Ni_3Al) are receiving attention due to their excellent high temperature behavior. These aluminides offer lower densities than conventional superalloys and have the potential to replace superalloys in jet engines [9]. One of the main drawbacks for these aluminides is their brittleness. If the grain size of these materials is reduced to the nanometer size range, then enhanced ductility at ambient temperatures may be expected [2].

Reactor Design

In designing the reactor, classical metallurgical engineering considerations of thermodynamics, reaction kinetics, momentum, heat and mass transfer were employed. The primary limiting design factors are: a) the injection of feed material into the hot zone, b) residence time of the feed material in the hot zone, c) heat transfer to the solid particulates, reactor walls and the outer electrode, d) plasma-particle interaction, and e) the nucleation, condensation and quenching of the products.

Under plasma conditions, the presence of charged particles and very steep temperature gradients in the boundary layer can have a substantial effect on the flow and temperature fields around the solid feed through its influence on the local fluid properties[15]. Pfender and other investigators [15] have performed fundamental studies on plasma-particle momentum and heat transfer. Several correlations regarding drag, heat and mass transfer coefficients under plasma conditions have been developed which are taken into account for the momentum, heat and mass transfer calculations.

In producing nanocrystalline metals, one of the most important aspects of processing is that the injected feed material should be exposed to high temperature for a sufficient period of time to ensure their complete vaporization. To increase the residence time of the entrained particles in the plasma arc, the length of the arc was increased by establishing it between a variable height electrode in the torch (which acts as the anode) and a "donut" shaped electrode which was mounted down from the torch.

To ensure the proper penetration of the feed material into the highly dense column of plasma arc, the feed powders were injected through an injection ring mounted on the torch with a high velocity of carrier gas and with a counter swirl direction relative to the plasma gas swirl.

Experimental Setup

A schematic of the experimental system is given in Figure 1. The central part of the setup is the reactor which is shown in Figure 2. The reactor is made of steel and can be divided into three zones. The first is the plasma jet zone, second is the extended flow zone and reaction zone, and third is the rapid quench zone. The hot zones are tubular in geometry and are designed to allow for the high temperatures generated by the plasma arc. Theoretical calculations were performed for the temperature and velocity profiles expected in the hot zone. An energy balance was performed to establish the reactor diameter and to identify the proper feed rates. The time for complete vaporization was calculated using kinetic equations. Knowing the estimated time for complete reaction, the length of the hot zone was determined to ensure complete vaporization for the particulate material.

The plasma arc was established between the vertically mounted torch (fitted with a linear actuator) which acted as an anode and a "donut" shaped cathode located in the central part of the hot zone. The distance between these two electrodes was adjusted to increase the arc length and for the proper attachment of the arc on the lower cathode. This lower electrode was held in place by a boron nitride ring. A schematic of the electrode is shown in Figure 3. This part was electrically isolated from the upper section. The "donut", which was made of copper, was intensively water cooled to avoid any film boiling. A mixture of argon and helium was used as plasma gas. The hot zone was lined with a graphite lining and insulated with graphite felt. All of these zones of the reactor were water cooled. Powders were injected into the reactor by a powder feeder through a graphite ring placed at the end of the plasma torch. Argon was used as carrier gas for the powder feeding. A two color pyrometer was

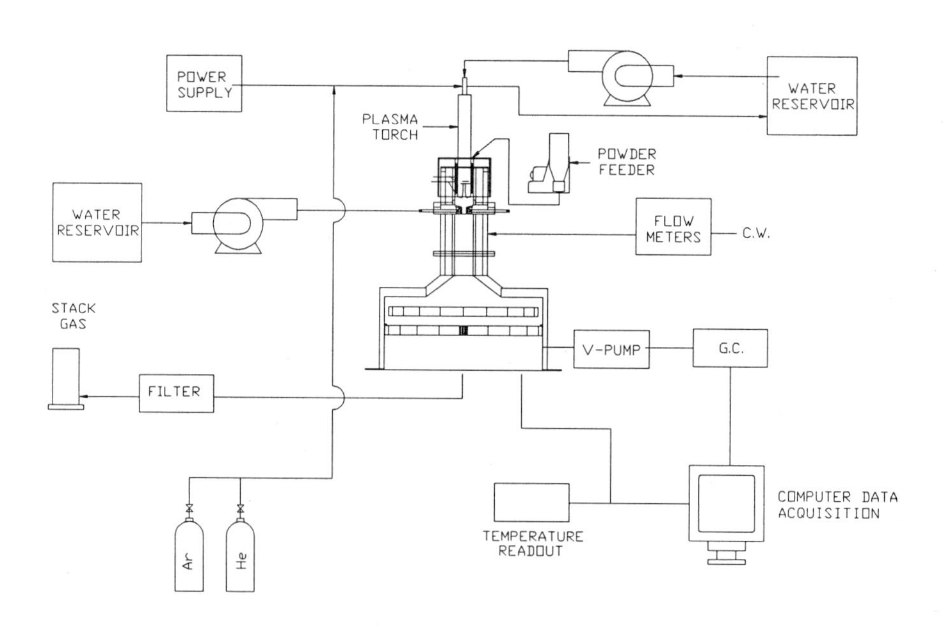

Figure 1 - Schematic of the experimental system.

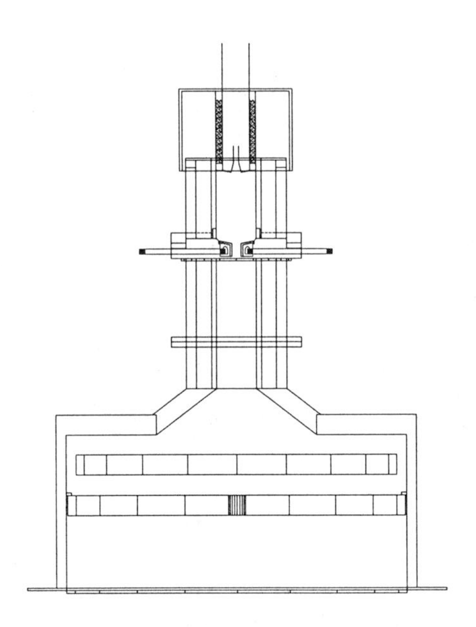

Figure 2 - Reactor schematic.

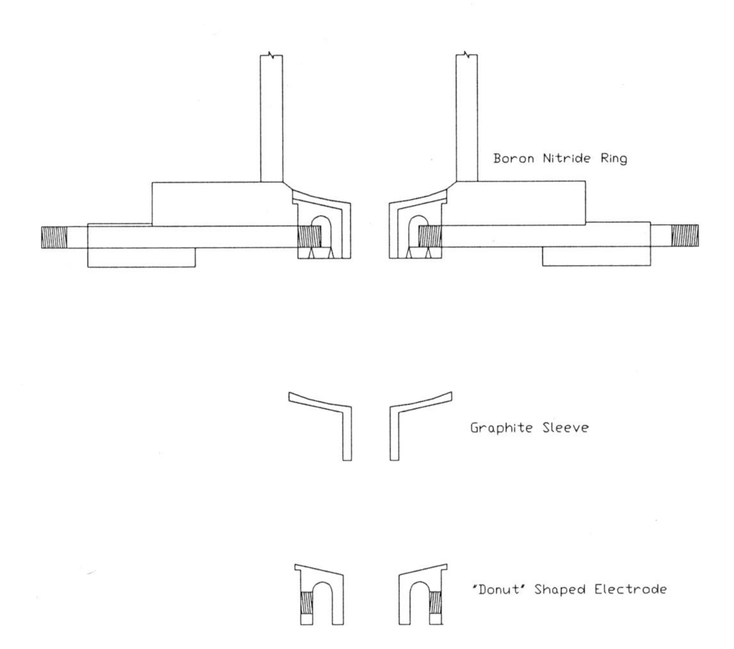

Figure 3 - Sketch of the lower electrode assembly.

used to measure temperature in the hot zone.

The hot section was attached to the quench box through a funnel section. The quench box contained two water cooled copper disks placed facing the discharge end of the hot zone. Most of the powder was collected from these disks. The disks were removable from the quench section by a hydraulic lift. Gases leaving the quench section were passed through a water collection system. Very fine particles entrained in the gaseous stream were captured in liquid and were recovered later by using appropriate settling reagents. A photograph of the experimental reactor is shown in Figure 4.

Figure 4 - Photograph of the reactor.

All of the information from the experiments such as temperature, gas, and water flow rates and other torch operating parameters was collected through an interface with a data acquisition system.

The general operating conditions for the experiments are given in Table 1.

Table 1: General Operating Conditions for the Experiments

Parameter	Value
Power	32 kW
Voltage	160 V
Current	200 A
Argon flow rate	3.0 SCFM
Helium flow rate	3.0 SCFM
Distance between the electrodes	4.5"
Powder feed rate	2.5-5 g/min

The first few runs were performed without any feed material to evaluate the operating behavior and performance of the experimental setup and to establish the proper running conditions for smooth operation especially from the arc stability point of view. In the initial experiments, the second section (extended flow zone) of the reactor was removed in order to observe the arc behavior under different operating conditions. Figure 5 shows the arc emerging from the lower electrode (cathode). After the initial open section experiments, an energy balance was performed at different electrode gaps from 2 inches to 7 inches.

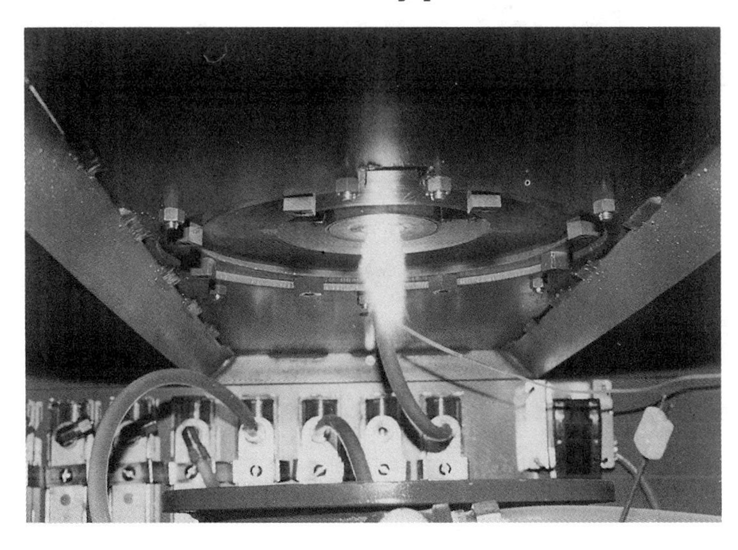

Figure 5 - Arc emerging from the lower electrode. (Ar-plasma gas)

Experimental Results

As mentioned earlier, the first stage of this work was to produce pure nanocrystalline metals. Aluminum, nickel and iron were chosen for these experiments. Reagent grade pure metal powders of different sizes were used as the feed material. The experimental conditions (shown in Table 1) were kept similar for all these metals. Aluminum powder (-325 mesh) was fed for twenty minutes at a rate of 5 g/min. A SEM micrograph of the feed material is shown in Figure 6. About half of the produced nanosize powder was collected from the copper disk in the quench box. About 30% was recovered from the water filter and the rest ended up in the funnel section and the top of the quench box. A SEM micrograph showing the nanosize aluminum powder is shown in Figure 7. Mostly the particles are in the range of 50-80 nm.

Next in the series of experiments was the nickel feed. Reagent grade nickel powder was used. Figure 8 shows the SEM micrograph of the nickel powder, which mainly consists of 5-20 micron sized particles. Nickel powder was fed at a rate of 3 g/min for 15 minutes. The produced powder was collected from the quench box and the water collection system. An SEM micrograph of the powder collected from the quench box is shown in Figure 9. Powder mainly consists of 50-80 nm sized particles.

Iron has higher melting and vaporization temperatures than

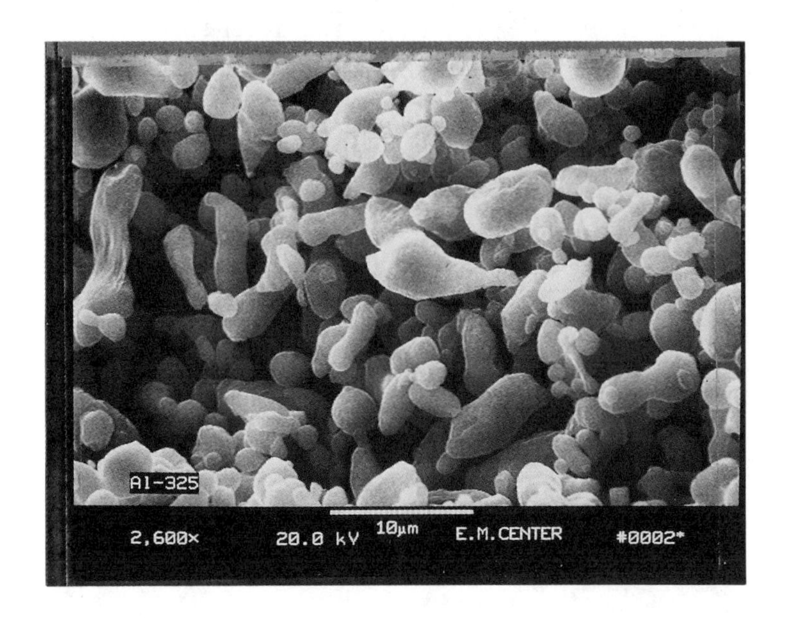

Figure 6 - SEM micrograph of the aluminum feed powder.

Figure 7 - SEM micrograph of the aluminum powder produced.

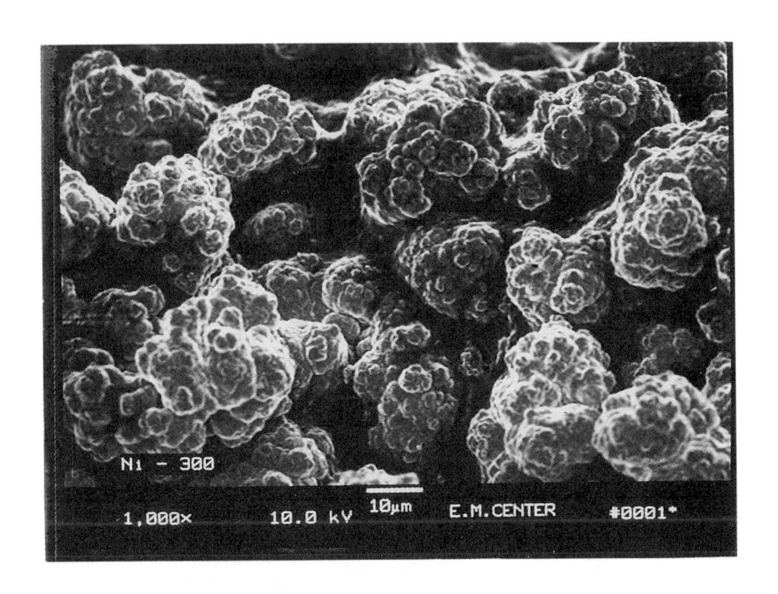

Figure 8 - SEM micrograph of the nickel feed powder.

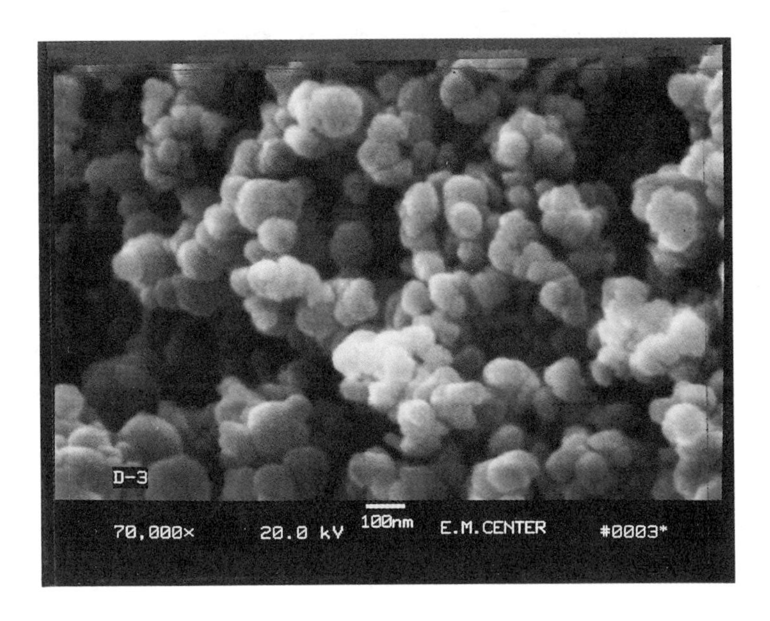

Figure 9 - SEM micrograph of the nickel powder produced.

the metals mentioned above. Reagent grade iron powder (-325 mesh) was fed into the plasma reactor at a rate of 2.5 g/min. The iron particles, after quenching were collected from the different areas of the quench box and the filter. About 30% were entrained in the gas leaving the quench box. Figure 10 shows the SEM micrograph for the powder collected from the quench box. The mean particle size in this micrograph is 48 nm.

Preliminary tests were done for the synthesis of nanosize intermetallics. For the synthesis of nickel aluminide, aluminum (-325 mesh) and nickel powder (-325 mesh) were used as reactants. Both the powders were mixed in the powder feeder with a molar ratio of (1:1). Powder was fed into the reactor at a feed rate of 2.5 g/min. X-ray diffraction results on the quench box powder showed the presence of Al, Ni, and NiAl phases. Synthesis and characterization of nanocrystalline intermetallics is underway.

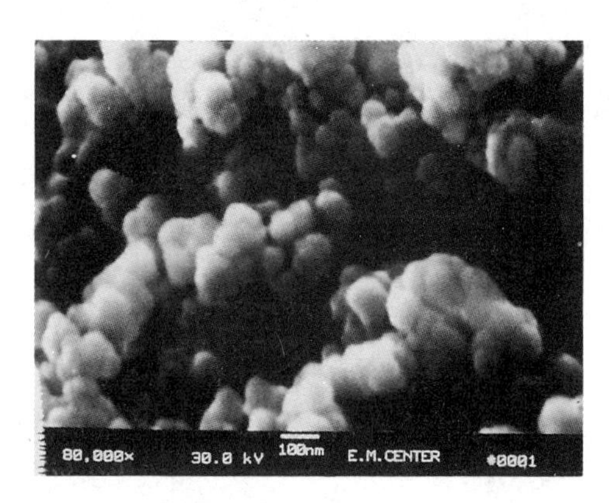

Figure 10 - SEM micrograph of the iron powder produced.

Conclusions

A novel thermal plasma reactor has been designed and built. The reactor provides for the development of an extended plasma flame by utilizing a variable height transferred arc to a donut shaped electrode. Nanocrystalline metals are produced in the thermal plasma reactor. The reactor appears to be capable of producing sufficient quantities of nano-sized materials for subsequent consolidation and analysis.

Acknowledgement

This research is supported by the U.S. Department of the Interior's Bureau of Mines under contract no. JO134035 through the Department of Energy's Idaho Field Office contract no. DE-AC07-76ID01570.

References

1. R. W. Siegal and H. Hahn, "Nanophase Materials," Current Trends in the Physics of Materials, World Scientific Pub. Co.,

Singapore, (1987), 403.

2. H. E. Schaefer and R. Wurschum, "Nanometre-sized Solids, Their Structure and Properties," J. Less Common Metals, 140(1988), 161-169.

3. F. H. Froes and C. Suryanarayana, "Nanocrystalline Metals for Structural Applications," J. of Metals, 6(1989), 12-17.

4. H. Gleiter, "Materials with Ultrafine Microstructures," Nanostructure Materials, 1(1992), 1-19.

5. R. W. Siegal and J. A. Eastman, "Synthesis, Characterization, and Properties of Nanophase Ceramics," Mat. Res. Soc. Symp. Proc. Vol. 132(1989), 3-14.

6. "Nanocrystalline Materials," Encyclopedia of Materials Science and Engineering, Suppl. Vol. 1, (R. W. Cahn ed.), Pergamon Press, 1988, 339-349.

7. R. W. Seigal, "Cluster Assembly of Nanophase Materials," Processing of Metals and Alloys, Vol. 15, (R. W. Kahn, ed.), VCH Verlagsgesellschaft, Weinheim, 1991.

8. C. Suryanarayana and F. H. Froes, "The Structure and Mechanical Properties of Metallic Nanocrystals," Met. Trans. A, 23A, 4(1992), 1071-1081.

9. F. H. Froes, et al., "Development, Technology Transfer, and Application of Advanced Aerospace Structural Materials," Structural Applications of Mechanical Alloying, F. H. Froes and J.J. deBarbadillo (eds.), ASM, Ohio, 1990.

10. R. M. Young, and E. Pfender, "Generation and Behavior of Fine Particles in Thermal Plasmas - A Review," Plasma Chem. & Plasma Proc., 5 (1) (1985), 1-37.

11. P. C. Kong and Y. C. Lau, "Plasma Synthesis of Ceramic Powders," Pure and Appl. Chem., 62 (9) (1990), 1809-1816.

12. P. R. Taylor and S.A. Pirzada, "Synthesis of Ceramic Carbide Powders in a Non-Transferred Arc Thermal Plasma Reactor", Thermal Plasma Applications in Materials and Metallurgical Processes, TMS, Warrendale, PA, 1992, pp. 249-268.

13. T. Harada, et al., "Synthesis of Ultrafine Powders of Nb-Al and Nb-Si Alloys by Using RF Plasma Reactor," 5th. Inter. Symp. Plasma Chemistry, 1981, 838-843.

14. C.H. Chou and J. Phillips, "Plasma Production of Metallic Nanoparticles," J. Mater. Res., Vol. 7(8), 1992, 2107-2113.

15. J. Feinman, ed., Plasma Technology in Metallurgical Processing, Iron & Steel Society, Inc, Warrendale, PA, 1987.

POWDER FEEDERS FOR PRODUCINGSTABLE LOW FLOW RATE SUSPENSIONS FROM COHESIVE POWDERS.

R.L. Stephens and B.J. Welch
Department of Chemical and Materials Engineering

J.S. McFeaters
Department of Mechanical Engineering
The University of Auckland
Private Bag 92019
Auckland
New Zealand

Abstract

Very fine powders are usually cohesive and difficult to feed reliably, especially at the low flow rates required for many laboratory-scale apparatus. As part of a thermal plasma synthesis project, it was necessary to develop a feeder capable of feeding pigment grade TiO_2 in an inert gas stream at feed rates of the order of 0.05 to 0.5 g/min. A vibrated fluidised bed for feeding pigment grade TiO_2 has been developed and results of trials using this feeder system are presented and discussed in terms of the satisfactory performance of this feeder for plasma synthesis reactions. In addition, a comprehensive literature review of this topic has been completed and an overview of the findings are presented. Finally, conclusions are drawn, from our results and findings of the literature review, on the suitability of various powder feeder designs for feeding cohesive powders.

Plasma Synthesis and Processing of Materials
Edited by Kamleshwar Upadhya

Introduction

The production of a uniform suspension of fine cohesive particulate material in a gas stream is crucial to the success of many applications of plasma processing specifically, and materials processing generally. A reduction in the primary particle diameter of the powder feedstock almost always increases the difficulty of producing a uniform suspension of the powder. This increase in difficulty occurs because surface interactions become the dominant feature determining the behaviour of the particles. It is important to note that the precise effect of the various factors will depend entirely on the system under study. For example, in the case of electrostatics, the tube material through which the suspension flows may significantly affect the results, so that no general conclusions may be drawn a priori [1].

A brief discussion of the underlying surface interaction phenomena affecting powder suspension formation can be found in section 2 of this paper. Many designs of powder feeders have been published in the literature. These designs have been categorised into one of five categories for this paper, and a discussion of the characteristics of each category of feeders, along with examples suitable for feeding cohesive powders, makes up section 3 of this paper. Specifications for a powder feeder for feeding submicron TiO_2 to an RF plasma torch for ceramic synthesis are then formulated in section 4. A vibrated fluidised bed, using elutriation from the fluidised bed as the feeding mechanism, was selected for development to meet the specifications, and the characteristics of this feeder are presented and discussed in section 5 of this paper. Finally, conclusions are drawn, from our results and findings of the literature review, on the suitability of various powder feeder designs for feeding cohesive powders.

Cohesive Forces

As the primary particle size decreases, the surface area to volume ratio of the particles increases and the relative effects of surface interactions increase. The underlying surface interaction phenomena that increase the difficulty of feeding ever finer powders are relatively well-known qualitatively but difficult to quantify. These include the effect of van der Waals forces, geometric factors, capillary forces, electrostatic forces, and gas absorption on the behaviour of the powder. The action of these underlying forces makes it almost inevitable that very fine (submicron) powders will be fed as agglomerates, with a larger apparent particle diameter than the primary particle diameter.

van der Waals Forces: The dominant interaction force between powders, and powders and the walls of the container in which they are stored, is the van der Waals force of attraction [1]. Van der Waals forces are the result of temporary fluctuations in the electronic field of a molecule which gives the molecule a temporary dipole character. This temporary situation will induce a change in neighbouring molecules, also rendering them temporarily dipolar. As a result of these temporary changes, and as a consequence of the general attraction between dipoles, the molecules and the particles composed of the molecules, attract each other.

Van der Waals forces act in both the gaseous and liquid environments, although the effect tends to be substantially reduced in liquids. Van der Waals forces are only noticeable between particles at separation distances of the order of a few molecular diameters [1], i.e. 0.2 to 1 nm, which implies that an environment in which contact between the particles is possible is a pre-requisite for the establishment of the force. The particle size where the van der Waals forces become significant is of the order of a few microns. Thus, powders made up of particles with a major dimension of less than about 5 to 10 μm, tend to become very cohesive due to van der Waals forces. Other forces can be superimposed on van der Waals forces resulting in an overall increase in cohesiveness.

Geometric Factors: The shape of the primary particles of a powder can substantially modify the flow properties of a powder in two ways. Firstly, surface asperities can interlock with each other resulting in an increase in cohesiveness. Interlocking tends to be important for particles larger than those considered here. Geometric factors can modify the influences of other forces by changing the contact area over which the molecular scale forces can be established. For example, asperities on the surface of particles are almost always greater than atomic scales so that the interfacial area between two particles is usually much less than first thought, which reduces the cohesiveness due to van der Waals forces. However, for porous particles or crystals with large flattened surface areas, the effect of the van der Waals forces may be larger than calculated based on a simple particle diameter, particularly if the particles exhibit cleavage along crystal planes because the fracture surface tends to be flat on an atomic scale.

Capillary Forces: At higher humidities ($\geq \sim 65\%$), capillary condensation of fluid in the gap between particles in close contact may occur, resulting in an attractive force component in addition to the van der Waals forces.

Electrostatic Forces: Powder processing and handling operations can dramatically alter the properties of a powder due to the development of electrostatic charging through contact electrification [2]. It is often more appropriate to describe such a contact electrification process as triboelectrification, as sliding or frictional contact is invariably involved, either between other particles or contact with walls of vessels that contain the powder.

The various factors affecting charge transfer are difficult to describe when insulating surfaces are involved in the process [2]. They include the need for donor/acceptor sites on the insulator surface, the need for accurately determined contact areas (geometric factors), the possibility of material as well as charge transfer during collisions, the kinetic nature of the charge transfer process where charge equilibrium is probably not reached, a possible reduction in the work function of smaller particles compared with a larger particle of the same material, and the effects of relative humidity.

Absorbed Gases: Physically absorbed gases increase the cohesion of powders, and the cohesion increases with an increase in gas pressure [3]. Their effect was found to be heavily dependent on which gases were used and the increase due to pressure was probably due to a measured linear increase in gas absorption with pressure.

Methods of Predicting Powder Behaviour: One of the more useful classification of the cohesiveness of a powder is the scheme developed by Geldart [4] for predicting the behaviour of fluidised bed systems. This scheme can be extended to applications such as vibratory or screw feeders to predict whether these feeders will produce a uniform suspension because the particle interactions occurring during fluidisation are the same as those occurring in other powder feeding systems. The scheme uses the difference between the particle and fluidising gas densities, and the primary particle diameter, to predict the powder's behaviour when it is used to form a fluidised bed. This scheme divides powders into four categories. The most easily recognizable features of the groups are [4]: powders in group A exhibit dense phase expansion after minimum fluidization and prior to the commencement of bubbling; those in group B bubble at the minimum fluidization velocity; those in group C are difficult to fluidize at all and those in group D can form stable spouted beds.

The powder categories of interest to this paper are groups A and C. Group A powders can be smoothly fluidised and generally have an average particle diameter of the order of 40 μm or larger. Powders which are in anyway cohesive belong in the group C category. They are typically extremely difficult to fluidise compared to group A powders, and generally have average particle diameters of less than 20 μm. As a general rule of thumb, if a powder is classified as a group C powder for fluidisation, then it is unlikely that a vibratory feeder or a screw feeder can be used to produce a satisfactory suspension from that powder due to the cohesive nature of the powder, which will cause clogging of the feeder and/or irregular feed rates.

Literature Review of Powder Feeders

There have been a wide range of powder feeder designs described in the open literature. These designs can be divided into one of five general design categories. These categories are based on the physical mechanism upon which each type of feeder is based. The five categories are:

- vibratory feeders,
- rotary valves and entrainment feeders,
- fluidised beds,
- liquid or slurry feeders, and
- the Wright system.

Within each category there will obviously be sub-categories which depend on differences as to how the feeder achieves the physical mechanism of feeding. For example, fluidized beds can produce a stream for feeding to a process by either elutriation from the bed surface or by removing a fraction of the powder through an orifice in the bed wall in the dense-phase region of the fluidized bed. It should be reiterated at this point that for very fine powders almost all of the powder feeders discussed will produce a suspension consisting of agglomerates of the primary particles, rather than a suspension in which there is negligible agglomeration of the primary particles.

Vibratory Feeders: Vibratory feeders typically use the vibration source to convey the powder to a point where it can then become entrained with a conveying gas to form the required powder suspension. Appropriately designed vibratory powder feeders can be used to produce high concentration suspensions suitable for laboratory-scale processes provided the powder is not too cohesive. Vibratory feeders also find wide application in pilot plant-scale and industrial-scale processes, particularly when particle sizes are greater than about 50 μm. Vibrations sources used in vibratory feeders are typically either mechanical, where motors fitted with eccentric cams are used to create vibration, electromechanical, where a solenoid is use to cyclically move a mass inducing a vibration, or pneumatic, where gas pressure is used to move a mass in much the same manner as for electromechanical sources. Frequencies vary between a few Hz and a few thousand Hz, with most systems running at around 100 to 200 Hz.

Several vibratory feeders have been developed for feeding cohesive powders. Knapp et al. [5] constructed a vibratory powder feeder to admit very small amounts of powder to pressurized gas streams (5 to 10 psig) at a controlled rate. They used their solenoid feeder to feed ultrafine talc ($3MgO.4SiO_2.H_2O$) and MgO. It is expected that the feeder would have been feeding agglomerates of the powder, rather than individual primary particles. An alternative powder feeder using a vibrating diaphragm was developed by Monroe [6]. In this design, instead of feeding the powder through a hole in the diaphragm, a gauze mesh cylinder filled with powder was attached to the diaphragm. The diaphragm was vibrated by a solenoid controlled by an autotransformer. The powders passed through the flexing gauze cylinder and were then entrained in a carrier gas. It is not clear what the primary particle size was. Reed [7] also used vibratory feeders for feeding a powders to a plasma torch and like Monroe, did not give any details of the particle size or feed rate.

Other examples of vibrating powder feeders capable of feeding very fine, and possibly mildly cohesive, powders that have been reported in the literature include the vibrating screen powder feeder as used by Humbert et al. [8] and a commercial vibratory powder feeder as used by Sakanaka et al. [9].

Rotary Valves and Entrainment Feeders: The underlying principle of these feeders is that the powder is delivered by a rotary motion and/or simply placed in a gas stream and the flow patterns or velocity of the entraining gas is such that the powders are entrained into the flow. Screw feeders are included in this category because they rely on a rotary motion to transport and meter the powder to the entrainment zone, and therefore control its delivery rate.

Screw feeders can only be used successfully with powders that exhibit a very slight propensity towards agglomeration. This is because the particles in the powder are forced into close contact with each other and the surfaces of the screw and bore wall. For example, Davies et al [10] found that the feed rate from their screw feeder was erratic, with either binding in the thread or clumping of the conveyed powder, unless the powders were greater than 50 μm without any fines. They also found that the reproducibility of the feed rates was very sensitive to the moisture content of the powders, with an increase in moisture increasing the cohesiveness of the powders, and therefore making the feed rates more erratic.

Slightly more cohesive powders can be fed using positive displacement entrainment feeders such as those used by Davies et al. [10,11], Huska and Clump [12], and Vissokov et al. [13]. The Davies et al. [10] feeder is a typical example of this class of feeder. The feeder consisted of a teflon piston set in a brass block and the block also had a 2 mm diameter gas injection line drilled such that the high velocity gas jet blew perpendicular to the movement of the piston. The powder was placed on the piston whose movement was controlled by a cam system. As the piston moved up, the gas jet entrained the fine powder exposed to the gas flow and then carried it to the plasma torch. It was found that the powder feed rate was linearly dependent on the piston rise rate, implying that the gas flow rate (0.5 to 2.0 l min^{-1}) was always sufficient to entrain all the powder. The maximum powder feed rate with 400 mesh (sub-38 μm) alumina was about 200 g h^{-1} (3.3 g min^{-1}) corresponding to a suspension concentration of 1.65 g l^{-1}. However, they note that this feeder would probably not be suitable for feeding agglomerative powders.

A very similar design concept to the piston feeders was used by Gullett and Gillis [15] and Burch et al. [16] for feeding more cohesive powders. In both designs, the position of the bed of powder is fixed and the point of gas entry is progressively moved into the powder bed so that the powder immediately in front of the gas injection point is entrained in the gas flow, and carried from the feeder. Gullett and Gillis [15] used their feeder to feed highly agglomerative calcium hydroxide ($Ca(OH)_2$) with a mean particle diameter of 1 to 3 μm. Data presented in the paper clearly showed that the powder suspension produced by this feeder consists primarily of agglomerates, rather than primary particles, due to the highly agglomerative nature of the calcium hydroxide. Interestingly, they designed this feeder to replace a vibrated fluidised bed system in an effort to minimise particle agglomeration by reducing the amount of particle-particle contact and thus produce a suspension of finer agglomerates. Burch et al. [16] developed a feeder to feed pulverized coal at very low flow rates (0.025 g min^{-1} to 0.14 g min^{-1}) at a maximum suspension concentration of 0.175 g l^{-1} with only small variations as a function of time. The short-time-scale stability of the feeder was evaluated indirectly, but indicated that the variability as a function of time was less than 3.0% provided the feeder was loaded with care to ensure uniform packing densities.

An alternative mechanism of delivering the powder to the entrainment point is to use a modified rotary valve. Giacobbe [17] designed a series of feeders for feeding industrial carbon black powder, which would be expected to have a mean particle size in the sub-micron size range, into a plasma reactor. Industrial carbon black powder is known to be highly cohesive. The third generation feeder was capable of producing a relatively wide range of suspension concentrations, from about 0.8 to 3.2 g l^{-1} which compare favourably with powder feeders described by Davies [18], at carbon dioxide flow rates of between 75 and 200 l min^{-1}. A vibrator was attached to the powder storage hopper to improve the powder flow properties into the rotary valves. Three different rotary valve designs were used to meter the powder from the hopper to an entrainment chamber. A commercially produced feeder, the GMD 60/2, utilising similar principles to those of the Giacobbe feeders, is available from Gericke AG [19]. The manufacturers claim that powders ranging from calcium stearate to corn starch to titanium dioxide can be fed at various flow rates using their feeder.

Davies [18] published a review of various powder feeding systems suitable for the production of concentrated powder suspensions at low flow rates. One of the various systems that he described was a circulating flow system which can be classified as an entrainment system. The principle of this system is to establish a large circulating volume of the suspension from which a small volume flow may be withdrawn for test purposes. The one potential drawback of this system is that it would be difficult to perform a mass balance on the system to determine the total powder flow rate over the period of powder feeding.

Fluidised Beds of Cohesive Powders: Fluidised beds are commonly used for feeding fine powders to experimental equipment, including plasma reactors. The fluidised beds can be operated in two modes depending on the suspension concentration required for the experiments. Dense phase feeding can be accomplished by extracting a suspension from a tapping in the side of the chamber containing the fluidised bed so that part of the dense phase bed is drawn off. Alternatively, the powder can be allowed to elutriate from the surface of the bed by entrainment of the smaller particles in the fluidising gas. The fluidising gas is then used as a carrier gas to convey the suspension to the experimental equipment. Elutriation invariably results in a lower suspension concentration than dense phase feeding.

Dense phase feeding of fine powders from a fluidised bed has been subject of several papers [20-22]. Chin et al. [20] used an annular fluidised bed in which the gas/solid suspension was withdrawn into a central pipe instead of using a side tapping. Copper concentrate particles with a density of 4010 kg m^{-3} and mean diameters of 64, 90 and 159 μm, were used with air as the fluidising gas. The solids loading ratios (142 g l^{-1} to 1535 g l^{-1}) achieved in these experiments appear extremely high compared to other methods of powder feeding. However, the particle size of these powders is relatively large and therefore the powders would be expected to be relatively free-flowing.

Geldart, Harnby and Wong [21] have investigated dense phase feeding from fluidised beds of cohesive powders as part of a larger study on the fluidisation of cohesive powders. They used a 4 mm diameter hole with a rounded entrance placed near the distributor to compare the efflux of free-flowing group A 70 μm alumina with group C 10 μm and 5 μm alumina. They found that the efflux rate of the group C powders was much lower than expected which they attributed to agglomeration of the primary particles resulting in larger secondary particles that can be shown theoretically to have a lower efflux rate. In support of this hypothesis, they calculated the agglomerate particle size by working back from the experimental results and using existing correlations for group A powders. Their analysis showed that the agglomerate size was larger near the bottom of the bed were it is expected that the compressive force due to the static head should have more effect and bring particles closer together allowing increased agglomeration due to increased interparticle forces.

The gas leaving the top of a fluidised bed carries entrained particles with it. Entrainment occurs in both single component and multicomponent systems. As the gas travels upwards from the bed surface, the flux of entrained solids decreases until a certain height above the bed surface, called the transport disengagement height (TDH), is reached. The flux of entrained solids remains essentially constant at any height above the TDH with any losses being due to mechanisms such as, for example, agglomeration of particles in the freeboard [21]. Elutriation is defined as the separation or removal of fines from a mixture, and this may occur either below or above the TDH.

It is important to note that entrainment, even when carefully measured, is inconsistent over short periods of time (minutes) [23], and poor reproducibility in excess of $\pm 30\%$ is typical [24]. Particle properties that are important are average size, size distribution, shape and density. The terminal velocity of a single particle is important, but clustering can occur, and with particles of mixed sizes, the flow phenomenon becomes more complex because different size particles tend to move at different velocities while exerting drag effects on each other. The maximum entrainment rate is governed by the saturation-carrying capacity of the gas stream under pneumatic transport conditions.

Generally entrainment increases with increasing gas velocity (entrainment $\propto u^{2.5 to 6}$), gas viscosity, gas density, bed diameter, and fines concentration [24]. Entrainment decreases for increasing disengagement height (to the TDH), particle density ($\propto u^{-4}$), particle size ($\propto u^{-2}$) and gravity. It has been observed that there is a sharp rise in entrainment rates from small diameter beds and a minimum in the entrainment rate with intermediate bed diameters. The sharp increase is thought to be associated with slugging within the bed while the minimum is thought to be due to gas channelling. For beds diameters greater than 8 to 10 cm, entrainment is thought to not be affected by bed diameter [24]. Also for fine solids, entrainment appears to insensitive to bed depth except at very shallow beds where entrance effects due to the gas distributor intrude. However, entrainment increases with deeper beds containing solids which fluidise poorly. This is due to the onset of severe channelling and slugging.

Geldart et al. [21] also examined the elutriation of cohesive powders from fluidised beds. They note that elutriation is a complex process and that the mechanisms are not well understood even for free-flowing solids. The phenomena include the ejection of solids into the freeboard by bubbles, hydrodynamic interparticle effects (shielding), collisions and momentum interchange, agglomeration, and wall effects. Elutriation is particularly affected by electrostatic effects. They found that contrary to expectations based on group A and B powders, the elutriation rates for cohesive powders are significantly lower than for free-flowing powders. They ascribed this primarily to be due to poorer fluidisation but agglomeration could increase the apparent particle size, or the more cohesive particles, on reaching the lower velocity region at the wall, may agglomerate with each other and/or stick to the wall thus disengaging more effectively. Visual observation indicated that there were fewer bubble eruptions in fluidised beds of cohesive powders. This observation agrees with the findings of Baron et al. [25] with regard to their proposed particle ejection mechanism of entrainment due to ejection of particles due to bubble collapse.

Fluidising very fine powders can prove to be extremely difficult because interparticle forces are relatively large compared to the fluidisation forces acting on the particles and can not be easily overcome. This means that the bed often behaves as a cohesive lump which will not break up to allow fluidisation. There are three primary methods for dealing with this problem. Firstly, agglomerate formation can be encouraged and then the minimum fluidisation velocity is that for the agglomerates. Secondly, larger particles can be used to create the fluidised bed and to this bed, a small mass fraction of fine powders is added. Thirdly, vibration can be used to break up the cohesive bed and to prevent it from re-forming. Occasionally these methods have been used in tandem to ensure adequate and reliable fluidisation [26].

Self-agglomeration was used by Morooka et al. [27] and Pacek and Nienow [28] to fluidise ultrafine powders. Morooka et al. examined the fluidity of submicron Ni, Si_3N_4, SiC, Al_2O_3, TiO_2, $CaCO_3$, and ZrO_2 using dry air or nitrogen as the fluidising gas. They found that except for $CaCO_3$ and ZrO_2, the powders were smoothly fluidised when the gas velocity exceeded an apparent minimum fluidisation velocity. All particles formed agglomerates during fluidisation, and the apparent minimum fluidisation velocity was a function of the agglomerate size that the submicron powders formed.

The equilibrium size of the agglomerates was determined by working backwards from experimental results and existing correlations for group A fluidised beds. It was found to be in the range of 70 to 700 μm. The agglomerates of the $CaCO_3$ and the ZrO_2 were found to be too cohesive to fluidise. $CaCO_3$ could be fluidised for short periods after the bed structure was broken up by vigorous shaking and an accompanying increase in fluidising gas velocity. Heating the bed during fluidisation of Si_3N_4 decreased the size of agglomerates by about 10 %, which implied that chemically adsorbed water was partly responsible for the fluidisation quality in this powder. Kusakabe et al. [29] observed similar agglomerate formation in fluidised beds formed from submicron powders at pressures down to 1 kPa.

Pacek and Nienow [28] found that they could fluidise a 94 % tungsten carbide/ 6 % cobalt hardmetal powder ($\rho \approx$ 14000 kg m^{-3}) with a mean particle size of around 4 μm could be fluidised using humidity controlled compressed air. They discussed the development of the fluidised bed from start up with unagglomerated powder. Initially the bed moved en masse up the inside of the containing vessel until the bed fractured as though under compression, typically through slip planes, and the bed collapsed. Part of the powder remained as a solid mass but part turned very rapidly into agglomerates which filled the channels between the solid masses. Gradually the agglomerates broke the solid masses down into more agglomerates and the bed ended up in two layers with large agglomerates up to 2 mm in diameter in the quiescent bottom layer with a layer of finer agglomerates above which were smoothly fluidised. Increasing the velocity through the bed broke down the larger agglomerates in the bottom layer so that the entire bed was smoothly fluidised. Subsequent fluidisation with agglomerates occurred in a manner typical of group B powders. Note that agglomeration is a dynamic equilibrium process which depends on interparticle forces for the formation of agglomerates, while disintegration of the existing agglomerates is caused by the shearing force due to bubble and interparticle movement [30].

The second of the three methods of improving the fluidisability of cohesive powders is the addition of larger particles in the bed. The larger particles are believed to act as turbulence promoters [30] and are also believed to help agglomerate break up [18]. On their own, the larger particles are usually classified as either group A or B powders. A small mass fraction of a group C powder can be added and the fluidised bed is operated such that the group C powders are elutriated from the bed. Cheremisinoff and Cheremisinoff [24] state that the entrainment rate of the group C powders is proportional to the concentration of group C powder in the bed. This statement is in agreement with the findings of Bronet and Boulos [31]. However Kono et al. [32,33] observed that when the mass fraction of group C reaches a "saturation point", agglomeration of the group C powders occurs and the group C agglomerates separate from the group A powder. It appears that the "saturation point" occurs when the fraction of group C powder exceeds that which can be consumed as a binder with the group A powders to form a weak structure of group A-C powders.

Examples of the use of fluidised beds that used larger particles to form the bed were can found in papaers by Morooka et al. [30], Boulos and co-workers [31,34] and Wu and Themelis [35]. In all cases the primary particle size of the group C powder was less than 5 μm.

The third method that can be used to improve the fluidisation of cohesive powders is the use of vibration to break up any large scale bed agglomeration. It is expected that this method would merely prevent large scale bed agglomeration thereby allowing fluidisation of agglomerates; it should not greatly modify the size distributions of the agglomerates that form since the energy intensity of the vibrations are generally low compared to, for example, ultrasonic vibration which is commonly used to break up agglomerates, particularly in slurry systems.

The most comprehensive study of the benefits of vibration for fluidisation of cohesive powders was performed by Mori et al. [36]. They first determined that the optimum angle of vibration was 45° to prevent channel formation and to smoothly fluidize the fine particles. They observed that a wide range of submicron particles, including Al_2O_3 (0.4 μm), TiO_2 (0.148 μm), SiO_2 (≤ 1 μm), and activated carbon with a mean particle size of less than 20 μm, could be fluidised at relatively low gas velocities in a manner similar to group A powders. $MgCO_3$ and $CaCO_3$ were fluidised but required more aggressive vibration and higher gas velocities to fluidise successfully. The higher velocity was required to fluidise the larger agglomerates that these powders formed. For example, $MgCO_3$ formed agglomerates larger than 1 mm. It was possible to identify an optimum frequency, evaluated on the basis of maximising the bed expansion, to obtain good fluidisation at low gas velocities.

Mori et al. also found that the elutriation rate of the fine particles could be easily controlled. For the alumina particles used in their study, they found that the elutriation rate increased with decreasing particle size down to 5 μm but the elutriation rate sharply decreased with further decreases in particle size. It is highly likely that this observation was due to the formation of larger agglomerates in the fluidized bed as the primary particle size decreased, and the higher terminal velocities of the larger agglomerates reduced the elutriation rates at a set gas velocity.

A fluidised bed used for powder feeding will become depleted over time if the bed is not continually replenished. Two methods of dealing with this problem have been used in the literature to deal with this problem. The first method used a very large bed so that bed depletion over the required time period is within an acceptable limit. However this alternative would result in very low solids to gas ratios because of the large amount of gas required to fluidise a large bed. This method would be acceptable for applications such as laser doppler anemometry measurements but not for materials processing applications where a concentrated suspension is required over a relatively long term. The second method is to use another class of powder feeder, such as a rotary valve [22] or a vibrated screw feeder [35], to continually replenish the fluidised bed.

Liquid or Slurry Feeders: An alternative to the problems of accurately metering solid flows entrained in a gas stream could be to use either dissolved salts of a metal in an aerosol form, or to use a slurry in which ultrafine ceramic powder would be suspended in a suitable liquid, and the solution or slurry is then nebulised to form an aerosol for feeding to the plasma torch. The liquid aerosol method has been used by several research groups [37-41] where an aerosol was formed using an ultrasonic nebuliser and the aerosol was then pneumatically transported to the plasma torch. This method also finds widespread use in spectrochemical analysis.

Slurry injection has received little or no attention in terms of a particle feeding mechanism for plasma reactors. This method was investigated briefly by Wu [42] for feeding pigment grade TiO_2, but he may have chosen the wrong option of trying to form the slurry aerosol in the plasma torch itself using a two-fluid spray nozzle, rather than forming the aerosol prior to the injection probe and then using pneumatic transport to carry the aerosol into the plasma reactor. This idea deserves more investigation before being discarded. The primary problem with this technique is to produce a slurry aerosol that is fine enough for the intended application. The selection of a suitable liquid is also important. For some systems that involve chemical reactions, it may be possible to choose a liquid that can also be used as a reactant. Some information on forming fine slurry aerosols can be found in work on slurry fuels for combustion systems.

The Wright System: Wright [43] came to the conclusion that since it is so difficult to achieve control over a loose aggregation of powder, the powder should be packed tightly into a cake from which the powder could be re-dispersed under controlled conditions. He found that, contrary to his expectations, almost any dust with particle sizes 90 % or more below 10 μm could be compacted into a solid cake, and provided the powder was dry, it could be readily re-dispersed by scraping the surface of the cake. The powders investigated by Wright ranged from coal dust and silica "flour" to aluminium powder and uranium dioxide, and he found that they were all equally easy to handle. The most important factor in this method is to ensure that the powder cake is uniformly packed/compressed so that the powder feed rate does not vary significantly with time.

The concept behind the Wright system was embraced by Davies [18] in the powder feeder he designed to feed kaolinite. Davies found that the original Wright system was limited in the maximum suspension concentration it could provide to only 0.3 kg m^{-3} when operated on kaolinite. He therefore produced a modified Wright system feeder which incorporated a high-speed cutting device in place of the rotary blade. Concentrated suspensions of up to 2 kg m^{-3} were able to be produced, and micrographs of the powder leaving the feeder showed that the original powder size (90% < 2 μm) had been restored.

Stream Splitting as an Option: A powder suspension is often more easily created in a larger scale powder feeder than those often used for laboratory experiments. Also, it is often desirable to split a powder suspension flow for feeding to multiple ports. Munz [44] therefore investigated stream splitting to investigate the splitting behaviour of a number of simple splitters, to determine which parameters influence the quality of split an in which way, and to define operating conditions under which an acceptable split may be obtained.

Munz found that a two-way splitter operated either singly or with two in tandem, to provide a four-way split, provides an acceptable split both on the long (minute) and short (millisecond) term. A four-way splitter where the split was accomplished at one level was found to be unacceptable. The best position for the splitter was found to be directly below the feeder compared to after a long length of conveying line, although he noted that this may have been due to the influence of suspension concentration variations induced by the use of two gentle 90° bends in the experimental equipment.

The authors' experience with conveying cohesive powders such as pigment-grade titania in laboratory-scale feeders, suggests that stream splitting will probably not be effective for these powders because the powders would impact and adhere on the leading edge within the splitter. The cohesiveness of the powders would cause more deposition from the suspension onto the adhered matter, leading to an eventual blockage of the splitter.

Specifications of a Powder Feeder for RF Plasma Synthesis

As part of a thermal plasma synthesis project, it was necessary to develop a feeder capable of feeding cohesive pigment-grade TiO_2 (primary particle size of approximately 0.2 μm) in an inert gas stream at feed rates of the order of 0.05 to 0.5 g min^{-1}.Residence time limitations in the RF plasma torch make it necessary to use fine powder with a maximum particle or agglomerate diameter of the order of 30 μm to ensure that complete vaporisation of the powdered reactant occurs [45]. Uniformity of the suspension concentration as a function of time is important for the control of stoichiometry for reactions occurring within the plasma torch.

The Vibrated Fluidised Bed

The powder feeder developed to satisfy the specifications described in the last section consisted of a hopper and rotary valve assembly which sat above a small 13.6 mm ID fluidised bed. The rotary valve was used to periodically replenish the fluidised bed. The hopper, rotary valve and fluidised bed were suspended from a frame which was vibrated using two pneumatic linear vibrators. The vibrating frame was isolated from its base by springs mounted on each corner of the frame. The vibration frequency used for all trials was about 90 Hz. A photograph of the feeder and vibrating frame is presented in Figure 1.

A fluidised bed was formed from cohesive Tiona AG anatase TiO_2. The TiO_2 powder was hand-sieved through a 420 μm sieve prior to use to remove any excessively large agglomerates. The agglomerates that passed through the sieve were not observed to form larger agglomerates again during storage or subsequent use in the rotary valve or the fluidised bed.

Figure 1: Photograph of the assembled fluidised bed feeder mounted in the vibrating stand.

Agglomerates were elutriated from bed, and exited through a valve tree assembly attached to the fluidised bed-containing vessel 120 mm above the distributor plate. The valve tree assembly was used to purge the feeder of air and to establish stable operation of the powder feeder prior to switching the suspension flow to the plasma torch. Powder not fed to the torch was collected in a plastic bag to enable an accurate mass balance to be performed at the end of each experiment.

It was originally intended that the rotary valve and fluidised bed would be independently calibrated so that the powder feed rate from the rotary valve could be easily matched to the powder feed rate from the fluidised bed. Calibration trials for the rotary valve showed that the powder feed rate through the feeder increased linearly as the valve rotation speed increased as shown in Figure 2. However, the powder fed per opening on the rotary valve decreased as the rotation speed increased. This was observed to be due to incomplete filling of each opening as the rotation speed increased. The standard deviation of the powder conveyed per opening also increased with increasing rotation speed. Obviously, if the opening remains under the inlet from the hopper for less time, then there is more chance of a momentary bridge influencing the results before it can be destroyed by the vibration imposed on the system.

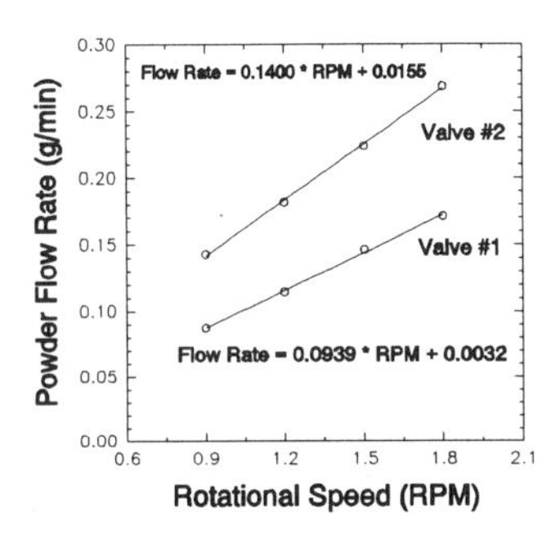

Figure 2: Calibration curves for the mark II rotary valve with valves no. 1 and 2.

The fluidised bed was initially characterised by filling the bed to a specified height and then weighing the bed to ± 0.01 g before and after the bed was fluidised for a set time. The times used were chosen to prevent more than a 15 mass% reduction in the mass of the bed during the trial. Results of experiments where slugging occurred in the bed were not recorded. The results of these first experiments are presented in Figure 3. The shapes of the curves presented in this diagram appear to have a very similar shape to data presented in Kunii and Levenspiel [46].

However, when the rotary valve and the fluidised beds were operated together as originally intended, it was observed that the fluidised bed depth slowly increased although the powder flow rates should have been matched using the data obtained from the individual calibration trials. The measured powder feed rate was also less then the expected 0.215 g min^{-1} for the experimental variables used. To explain this behaviour, it was necessary to perform long-term stability trials, in which the powder efflux rates during contiguous three minute periods were measured.

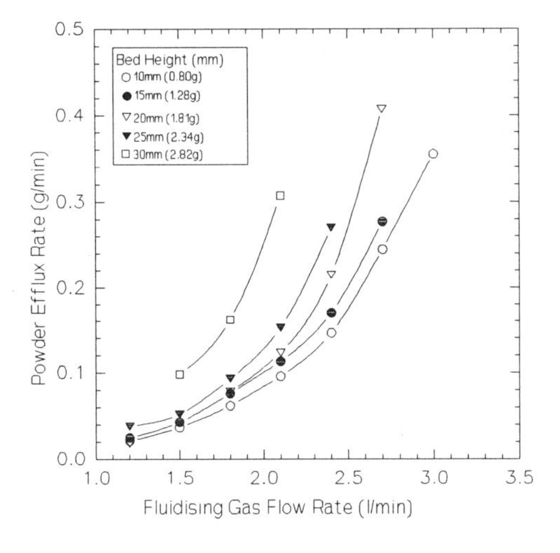

Figure 3: Calibration curves for the 13.6 mm diameter fluidised bed for sieved (sub-420 μm) Tiona AG anatase TiO_2.

The results of the powder efflux rate measurements are presented in Figure 4. In all cases, the powder efflux rate initially decreased sharply to a value much lower than that expected on the basis of the short-term efflux rate trials. The powder efflux rate then slowly increased for times longer than about six minutes. These observations can be explained as follows. Initially, the powder flow rate is high because of the high fraction of fines present in the fresh powder loaded into the bed prior to the experiment. The powder feed rate then falls to what should be a steady state flow rate.However, the mass of powder in the bed slowly increases, causing the top of the fluidised bed to rise closer to the outlet port. This effectively reduces the bed freeboard leading to an increase in the elutriation rate. Thus the powder efflux rate slowly increases.

It was considered that the observed changes in the powder efflux rates were unacceptably large for powder synthesis experiments, since the reaction stoichiometry would vary too widely during the course of a synthesis experiment.

In an attempt to overcome this variability, glass beads were added to the bed. It was hoped that a bimodal size distribution, with respect to fluidisation, could be created. In this case, the lighter/smaller titania agglomerates would be elutriated from the bed, while the larger/heavier glass beads would remain in the bed. It was later determined that this hypothesis would only be valid above the transport disengaging height (TDH), but the dimensions of the fluidised bed used made it impossible to operate above the TDH. Problems with operating below the TDH were further exacerbated by the slugging, rather than bubbling, behaviour in the fluidised bed under some operating conditions.

The results of the powder efflux rate experiments at the conditions later adopted for powder synthesis experiments are presented in Figure 5. The results show that the mass fed per three minute interval varies widely, but after the first three minutes, the average powder efflux rate only increases slowly. The large variations observed in the results were due to transitions from slugging flow to bubbly flow and back again during the course of an experiment, and the inherent variability of elutriation rates [24].

The averaged powder efflux rate, calculated disregarding the results for the first three minutes for each set of experiment, was 0.145 g min^{-1}. It was observed that the bed height increased during the experiment, and

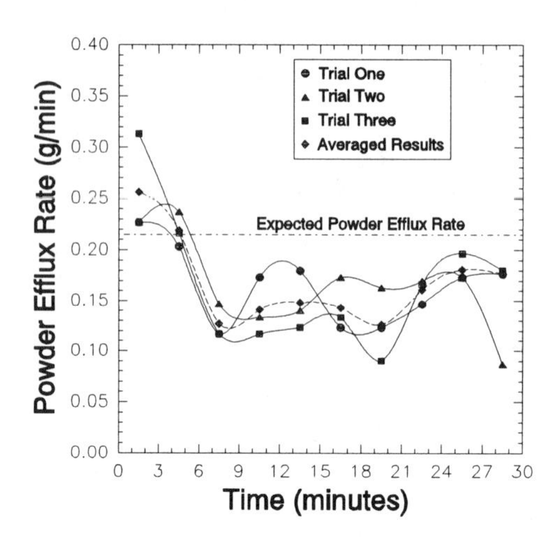

Figure 4: Powder efflux rates for the 13.6 mm diameter fluidised bed for a sieved 2 g (sub-420 μm) TiO_2 bed, 2.4 l min^{-1} Ar fluidising gas, and 1.5 RPM rotary valve speed.

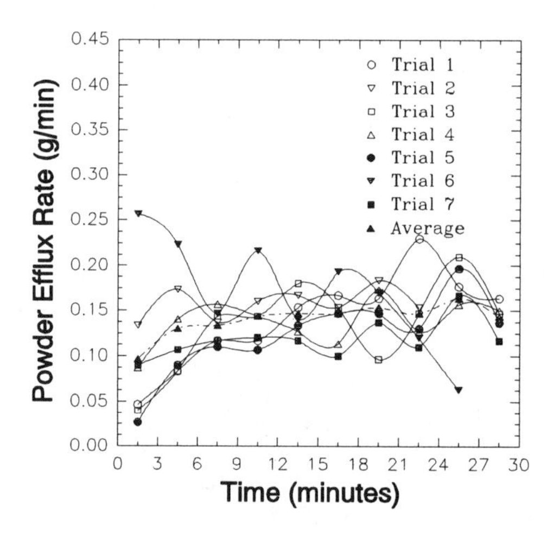

Figure 5: Powder efflux rates for the 13.6 mm diameter fluidised bed for sieved (sub-420 μm) TiO_2, 5.7 g glass bead bed, 3 l min^{-1} Ar fluidising gas, and 1.5 RPM rotary valve speed.

this was reflected in a steady slow increase in the average powder efflux rate as the experiment time progressed. The increase in bed depth was attributed to the formation of large agglomerates in the bed which were statistically unable to be elutriated from the bed.

The conditions under which the data presented in Figure 5 were generated were adopted for powder synthesis experiments. The presence of large, unreacted TiO_2 agglomerates was consistently observed in the products of these experiments. The agglomerates were too large to be vaporised in the plasma torch given the limited residence time in the hottest regions of the plasma. The agglomerates were formed because of the cohesive nature of the powder, and the nature of particle-particle contact within the feeder. It is considered that, while the feeder may be suitable for experiments in which agglomerate production is not a concern, the feeder used was unable to fully satisfy the required specifications.

The most suitable feeder for this project would have been a Wright system feeder, as the cutting process effectively destroys any agglomerate structure that may have been present in the powder compact. Preliminary experiments with a cutter similar to that used by Davies [18] showed that a compressed cake of TiO_2 could be redispersed with ease.

Summary

Very fine powders are usually cohesive and difficult to feed reliably, especially at the low feed rates required for many laboratory-scale apparatus. The underlying surface interaction phenomena that cause the cohesiveness of very fine powders are relatively well-known qualitatively but difficult to quantify. The phenomena discussed include the effects on the behaviour of the powder of van der Waals forces, geometric factors, capillary forces, electrostatic forces, and gas absorption. Van der Waals forces are usually the dominant cohesive force, although other forces can be superimposed on the van der Waals forces.

Powder feeder designs that have appeared in the literature can be divided into one of five general categories depending on the physical mechanism upon which each type of feeder is based. The five categories are vibratory feeders, rotary valves and entrainment feeders, fluidised beds, liquid or slurry feeders, and the Wright system. With the exception of the Wright system, all the powder feeders discussed will produce a suspension consisting of agglomerates rather than a suspension in which there is negligible agglomeration of the primary particles.

A specification was developed for a powder feeder for feeding pigment-grade TiO_2 to a plasma torch for use as a reactant in a ceramic powder synthesis project. A vibrated fluidised bed was developed and characterised as a possible suitable powder feeder to meet the specifications developed in this paper. Ceramic powders synthesised using this feeder suffered from heavy contamination from unreacted TiO_2. The unreacted TiO_2 was present in the form of large agglomerates that formed in the powder feeder and were unable to be vaporised, and then react, in the plasma torch. On the basis of these results, it can be concluded that the vibrated fluidised bed did not fully meet its design specifications. A more suitable classification of powder feeder for development to meet specifications such as those formulated for the application described in this paper would have been a Wright system powder feeder [18,43], as it should produce a suspension relatively free of large TiO_2 agglomerates.

Acknowledgments

This work was partially supported by DSIR Chemistry Contract No. UV/CD/3/1.

References

[1] J. Visser, "Van der Waals And Other Cohesive Forces Affecting Powder Fluidization", Powder Technol., 58, 1-10 (1989)

[2] A.G. Bailey, "Electrostatic Phenomena During Powder Handling", Powder Technol., 37, 71-85 (1984)

[3] H.W. Piepers, E.J. Cottaar, A.H.M. Verkooijen and K. Rietema, "Effects of Pressure and Type of Gas on Particle-Particle Interaction and the Consequences for Gas-Solid Fluidization Behaviour", Powder Technol., 37, 55-70 (1984)

[4] D. Geldart, "Types of Gas Fluidization", Powder Technol., 7, 285-292 (1973)

[5] M.R. Knapp, R.H. Arendt, R.A. Giddings and C.W. Krystyniak, "A Controllable Gas Stream Powder Feeder", Powder Technol., 14, 185-186 (1976)

[6] C.W. Marynowski and A.G. Monroe, "R-F Generation of Thermal Plasmas", in Proc. Int. Symp. High Temp. Technol., IUPAC, Pacific Grove, Calif., 67-84 and 525-540, 8-11 Sept. (1963) (Published by Butterworths 1964)

[7] T.B. Reed, "Growth Of Refractory Crystals Using The Induction Plasma Torch", J. Appl. Phys., 32, 12, 2534-2535 (1961)

[8] P. Humbert, D. Morvan, J.F. Campion, P. Jolivet and J. Amouroux, "Refining and Nitriding Si and Ti with a Plasma Torch", in Z.A. Munir and J.B. Holt (eds.), "Combustion and Plasma Synthesis of High-Temperature Materials", VCH Publ., New York, 454-469 (1990)

[9] K. Sakanaka, A. Motoe, T. Tsunoda, T. Kameyama and K. Fukuda, "Synthesis of Ultrafine Tungsten Carbide Powders with a Dual R.F. Thermal Plasma System", Proc. 9th Int. Symp. Plasma Chem., Pugnochiuso, Italy, Sept. 3-8, 882-887 (1989)

[10] G.J. Davies, R.M. Jervis and G. Thursfield, "Dispensers for Feeding Powders into Reactive Gaseous Environments", J. Phys. E., 3, 666-667 (1970)

[11] G.J. Davies, R.M. Jervis and G. Thursfield, "Processing New Zealand Titaniferous Sands In An Induction-Coupled Plasma Torch", N.Z. J. Sci., 13, 468-481 (1970)

[12] P.A. Huska and C.W. Clump, "Decomposition Of Molybdenum Disulphide In An Induction-Coupled Argon Plasma", Ind. Eng. Chem. Proc. Des. Develop., 6, 2, 238-244 (1967)

[13] G.P. Vissokov, K.D. Manolova and L.B. Brakalov, "Chemical Preparation Of Ultra-fine Aluminium Oxide By Electric Arc Plasma", J. Mater. Sci. Letters, 16, 1716-1719 (1981)

[14] P. Kong, T.T. Huang and E. Pfender, "Synthesis of Ultrafine Silicon Carbide Powders in Thermal Arc Plasmas", IEEE Trans. Plasma Sci., PS-14, 4, 357-369 (1986)

[15] B.K. Gullett and G.R. Gillis, "Low Flow Rate Laboratory Feeders for Agglomerative Particles", Powder Technol., 52, 257-260 (1987)

[16] T.E. Burch, R.B. Conway and W.Y. Chen, "A Practical Pulverised Coal Feeder for Bench-Scale Combustion Requiring Low Feed Rates", Rev. Sci. Instrum., 62(2), 480-483 (1991)

[17] F.W. Giacobbe, "Advanced Powder Feeding Device for use in Gas/Solid Plasma Synthesis and Processing Applications", Mat. Res. Soc. Symp. Proc., 98, 405-416 (1987)

[18] T.W. Davies, "The Production of Concentrated Powder Suspensions at Low Flow Rates", Powder Tech., 42, 249-253 (1985)

[19] Installation and Service Instructions, Gericke GMD 60/2 Powder Feeder, Althardstrasse 120, CH-8105 Regensdorf, Switzerland (1985)

[20] E.J. Chin, R.J. Munz and J.R. Grace, "Dense Phase Powder Feeding From An Annular Fluidized Bed", Powder Technol., 25, 191-202 (1980)

[21] D. Geldart, N. Harnby and A.C. Wong, "Fluidization of Cohesive Powders", Powder Technol., 37, 25-37 (1984)

[22] F.R.A. Jorgensen and T.M. Turner, "Technical Note on the Metering of Fine Powders at Low Flowrates", Proc. Australas. Inst. Min. Metall., 265, 41-43 (1978)

[23] R.H. Perry and D.W. Green, "Perry's Chemical Engineers Handbook", 6th ed., McGraw-Hill, New York, p20-62 (1984)

[24] N.P. Cheremisinoff and P.N. Cheremisinoff, "Entrainment Correlations", in N.P. Cheremisinoff (ed.), "Solids and Gas-Solids Flow", Vol. 4, "Encyclopedia of Fluid Mechanics", Gulf Pub. Co., Houston, pp1043-1062 (1986)

[25] T. Baron, C.L. Briens, P. Galtier and M.A. Bergougnou, "Effect of Bed Height on Particle Entrainment From Gas-Fluidized Beds", Powder Technol., 63, 149-156 (1990)

[26] Reference No. 5 in Reference [18]

[27] S. Morooka, K. Kusakabe, A. Kobata and Y. Kato, "Fluidization State Of Ultrafine Powders", J. Chem. Eng. Jap., 21, 1, 41-46 (1988)

[28] A.W. Pacek and A.W. Nienow, "Fluidisation of Fine and Very Dense Hardmetal Powders", Powder Technol., 60, 145-158 (1990)

[29] K. Kusakabe, T. Kuriyama and S. Morooka, "Fluidization of Fine Particles at Reduced Pressure", Powder Technol., 58, 125-130 (1989)

[30] S. Morooka, T. Okubo and K. Kusakabe, "Recent Work on Fluidized Bed Processing of Fine Particles as Advanced Materials", Powder Technol., 63, 105-112 (1990)

[31] M.S. Bronet and M.I. Boulos, "Particle Slip Velocities in Air Jets Under Atmospheric Pressure and Soft Vacuum Conditions", Can. J. Chem. Eng., 68, 353-359 (1990)

[32] H.O. Kono, S. Chiba, T. Ells and M. Suzuki, "Characterization of the Emulsion Phase in Fine Particle Fluidized Beds", Powder Technol., 48, 51-58 (1986)

[33] H.O. Kono, C.C. Huang, E. Morimoto, T. Nakayama and T. Hikosaka, "Segregation and Agglomeration of Type C Powders from Homogeneously Aerated Type A-C Powder Mixtures During Fluidization", Powder Technol., 53, 163-168 (1987)

[34] J. Lesinski and M.I. Boulos, "Laser Doppler Anemometry under Plasma Conditions. Part I. Measurements in a D.C. Plasma Jet", Plasma Chem. Plasma Proc., 8, 2, 113-132 (1988)

[35] L. Wu and N.J. Themelis, "The Flash Reduction of Electric Arc Furnace Dusts", JOM, 44, 1, 35-39 (1992)

[36] S. Mori, A. Yamamoto, S. Iwata, T. Haruta, I. Yamada and E. Mizutani, "Vibro-Fluidization of Group-C Particles and its Industrial Applications", AIChE Symp. Series No. 276, v86, 88-94 (1989)

[37] M. Kagawa, M. Kikuchi and Y. Syono, "Stability Of Ultrafine Tetragonal ZrO_2 Coprecipitated With Al_2O_3 By The Spray-ICP Technique", J. Am. Ceram. Soc., 66, 11, 751-754 (1983)

[38] X.W. Wang, H.H. Zhong and R.L. Snyder, "RF Plasma Aerosol Deposition of Superconductive $YBa_2Cu_3O_{7-\delta}$ Films at Atmospheric Pressure", Appl. Phys. Lett. 57, 15, 1581-1583 (1990)

[39] Y.C. Lau, P. Kong and E. Pfender, "Plasma Synthesis Of Ceramic Powders By Injection Of Liquid Precursors", Proc. 8th Int. Symp. Plasma Chem., Tokyo, Japan, 2381 (1987)

[40] Y.C. Lau, P.C. Kong and E. Pfender, "Synthesis Of Zirconia Powders In An RF Plasma By Injection Of Inorganic Liquid Precursors", 1st Int. Conf. on Ceramic Powder Processing Sci. (1987)

[41] P.C. Kong and E. Pfender, "Plasma Synthesis of Fine Powders by Counter-Flow Liquid Injection", in Z.A. Munir and J.B. Holt (eds.), "Combustion and Plasma Synthesis of High-Temperature Materials", VCH Publ., New York, 420-430 (1990)

[42] M.K. Wu, Personal Communication, October 1990.

[43] B.M. Wright, "A New Dust-Feed Mechanism", J. Sci. Inst., 27, 12-15 (1950)

[44] R.J. Munz, "Stream Splitting For Plasma Reactor Feeding", Powder Technol., 25, 79-84 (1980)

[45] M.K. Wu, "Aspects of Heat Transfer to Particles in Thermal Plasma Processing", PhD Thesis, The University of Auckland, July 1991

[46] D. Kunii and O. Levenspiel, "Fluidization Engineering", Robert E. Krieger Publishing Co., NY, pp302-325 (1977)

Author Index